T0207606

Springer Undergraduate Mathematics Series

SUMS Readings

SUMS Readings is a collection of books that provides students with opportunities to deepen understanding and broaden horizons. Aimed mainly at undergraduates, the series is intended for books that do not fit the classical textbook format, from leisurely-yet-rigorous introductions to topics of wide interest, to presentations of specialised topics that are not commonly taught. Its books may be read in parallel with undergraduate studies, as supplementary reading for specific courses, background reading for undergraduate projects, or out of sheer intellectual curiosity. The emphasis of the series is on novelty, accessibility and clarity of exposition, as well as self-study with easy-to-follow examples and solved exercises.

Ignacio Zalduendo

Calculus off the Beaten Path

A Journey Through Its Fundamental Ideas

Ignacio Zalduendo
Department of Mathematics
Torcuato di Tella University
Buenos Aires, Argentina

ISSN 1615-2085 ISSN 2197-4144 (electronic)
Springer Undergraduate Mathematics Series
ISSN 2730-5813 ISSN 2730-5821 (electronic)
SUMS Readings
ISBN 978-3-031-15764-6 ISBN 978-3-031-15765-3 (eBook)
https://doi.org/10.1007/978-3-031-15765-3

Mathematics Subject Classification: 26-01, 26A06, 40-01, 26Dxx

This Springer imprint is published by the registered company Springer Nature Switzerland AG
The registered company address is: Gewerbestrasse 11, 6330 Cham, Switzerland

Cálculo para Iñaki

Preface

After the publication of *Matemática para Iñaki* [13], I proposed to keep the promise expressed in its Prologue and write *Cálculo para Iñaki.* My intention was to give a reasoned, very accessible, and colloquial explanation of the main ideas of calculus, with some historical references, and centered on applications. Having taught calculus courses for over 40 years, I know that some aspects of the matter can be hopelessly arid and boring to students and teacher alike. Thus I also proposed to write a book emphasizing what to me are the more conceptually important aspects, and interesting applications, leaving aside—whenever possible—the technicalities and the purely computational. It soon became apparent that I had led myself into a complicated situation. The intentions I expressed above resulted at times incongruent and very difficult to reconcile.

The fact is that many of the interesting applications which I insisted on including (such as the Basel problem and the sum of Gregory's series) required more and deeper concepts, slowly distancing me from my original purpose of extreme accessibility. Thus, this is not the very elementary book that I set out to write, but rather the best I could do with non-elementary subject matter. However, I have strived for clarity and colloquiality, and in the end, I am happy with both the content and the tone of the text.

And so, with a different title and in another language, here it is. To show the spirit in which the book is written, perhaps it is convenient to list here some of the topics and applications which I did not want to leave out, and which are not commonly included in calculus courses:

- A construction of the real numbers
- Riemann's series theorem (rearrangement theorem)
- Proofs of the irrationality of $\sqrt{2}$, e and π
- Pythagorean triples
- The concept of limit in Ancient Greece
- Snell's law and the Brachistochrone
- Buffon's needle
- Growth of the harmonic series
- Gregory's series
- Stirling's formula
- Curvature

- Convexity
- Random walk and the bell curve
- The isoperimetric inequality
- Classical inequalities (AG, Jensen, Young)
- The Basel problem
- Density functions, barycenter, and expectation
- Pappus' theorem
- *The Method* of Archimedes
- The catenary
- The Gamma function

The educated reader may notice varying levels of informality and formality in different parts of the text. My personal inclination is more towards informality: I would much rather be accused of mathematical incompleteness than of lack of expository clarity. I will strengthen the hypotheses if this does away with an inessential technicality. However, at times, formality has permitted the discussion of important notions. For example, summing well Gregory's series gave me the opportunity to talk about uniform convergence, which is much more important than Gregory's series.

I am far from being an expert on the history of mathematics. But on re-reading some of the older sources for this book, I could not help but think that many of the underlying ideas of Calculus have been developing for 2400 years, certainly since before the time of Archimedes, although they come of age in the XVIIth Century. I have tried to point out the origin of some of these ideas in the text, without pretending that this is a history book.

Finally, my personal views on some of the subject matter included here, and which need not be shared by others, have also shaped the text. Among them, I must confess the following: Taylor polynomials of order one and two seem to me the more important, just as the first and second derivatives are those with immediate applicability and a clear geometrical significance. Integration is an area where I do not find formality particularly useful. Although I do address the difficulties of defining the integral, and I refer to Riemann's and Lebesgue's definitions, I adopt the intuitive idea of integral as the area under a continuous curve which was so productive until the XIXth Century. Also, I find inequalities are central to analysis, and I included a chapter discussing a few of them.

This book was transformed from illegible manuscript to elegant LaTeXtext by Pablo Sanches. I have received valuable suggestions and comments from Federico Poncio, Lara Sánchez Peña, Guillermo Ranea, Vicky Venuti, Damián Pinasco, Angelines Prieto, and Maite Fernández Unzueta. Finally, Robinson dos Santos has been a kind and understanding Editor to this rather stubborn author. To all, my deepest gratitude.

Buenos Aires, Argentina
June 10, 2022

Ignacio Zalduendo

Contents

Introduction

2400 Years of Calculus

In the Vth Century BCE, a tremor shook the foundations of Greek mathematics: the diagonal of a square turned out to be incommensurable with one of its sides. No unit of measure—no matter how small—would fit an integer number of times in one and the other. Mathematicians began to discern that rational numbers were insufficient to describe the lines and curves which appeared so clear to their intuition. Eudoxus then produces a new theory of proportions to compare incommensurable magnitudes. His ideas put him elbow-to-elbow with Richard Dedekind, who, with his "cuts," formalizes in 1872 the notion of real number which Eudoxus was seeking 2200 years earlier.

Eudoxus and Archimedes used successfully a notion of limit, which was formally defined in the XIXth Century. They calculated, using an idea analogous to the Riemann integral, areas of sections of parabolas and the volume of spheres. Archimedes explains in his work *The Method* the way in which he glimpsed some of his results. His method is very similar to the one used by Bonaventura Cavalieri in 1635. But Archimedes' work had been lost in the XIIIth Century and would not be rediscovered until 1906. The Hindu mathematician Madhava of Sangamagrama sums in 1400 Gregory's series, which the Englishman would sum in 1668.

What I mean by all this is that some of the fundamental ideas of calculus are as old as mathematics. And these ideas, methods, and results have appeared with greater or lesser clarity, with more or less transcendence, in the minds and in the works of many mathematicians, separated by thousands of years and thousands of kilometers. Eudoxus and Dedekind would have understood each other in 5 minutes.

But what is Calculus? Perhaps the leitmotif common to most tools, methods, and results which form part of the Calculus is the use of magnitudes which are, while not "infinitely small," at least "as small as required," usually in some process calling for smaller and smaller magnitudes.

An example: suppose we want to find the area between a and b under the curve

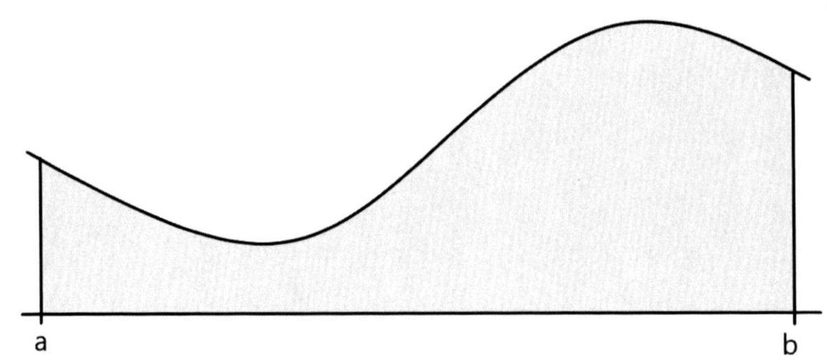

The problem: our units of area measure (cm^2, m^2,...) are rectangular and do not fit well against the curved boundary of what we want to measure. This makes comparison difficult.

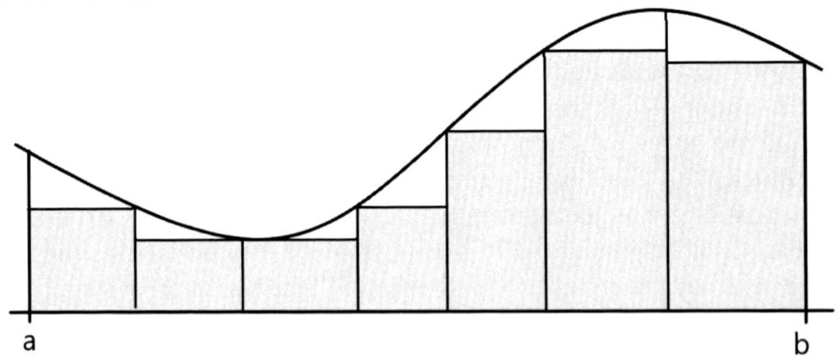

The solution: consider rectangles with base "as small as required," which will then adjust better and better to the area under the curve.

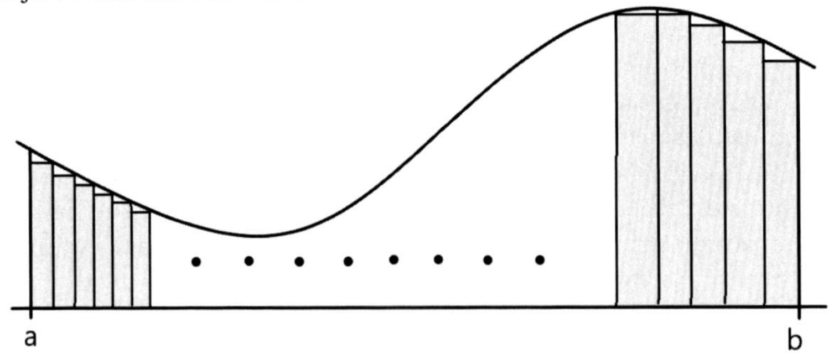

Another example: we want to find the line tangent to this curve at the point P.

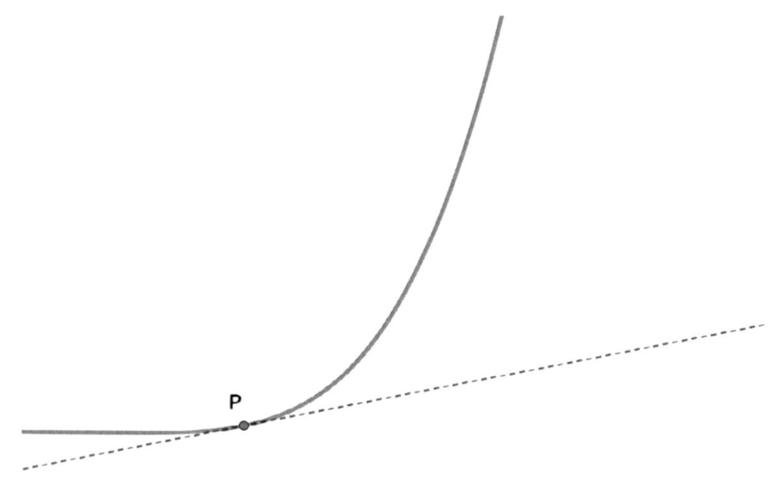

The problem: we know how to find lines passing through two points, but here we only know it passes through P (and we have an intuitive idea of which line we are referring to).

The solution: consider lines through P and another point Q on the curve, but then move Q to make it "as close as required" to P. These lines will be closer and closer to the tangent line.

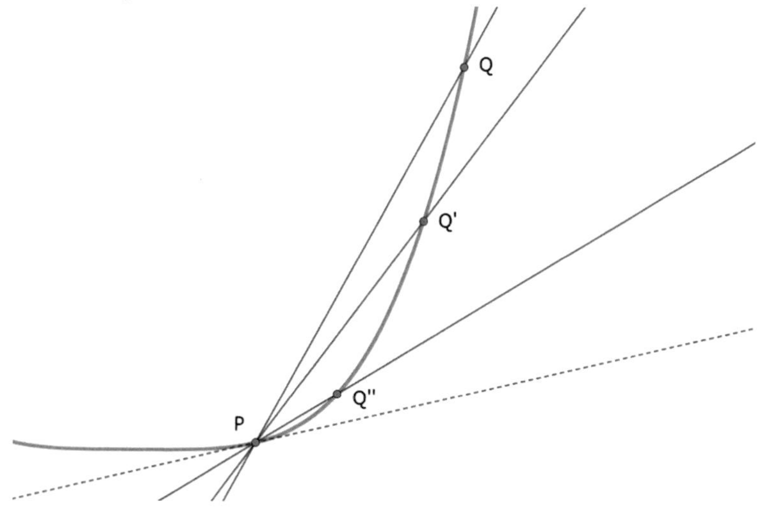

The first problem originates the part of calculus known as "integral calculus," and the second the part is known as "differential calculus." As we have said, integral calculus has its origins in Eudoxus and Archimedes. Differential calculus originates closer to the XVIIth Century with Fermat, Descartes, and others who wished to solve geometric problems or problems involving maxima and minima.

But in the XVIIth Century, with Newton and Leibniz, something extraordinary happens: integral and differential calculus are recognized as two sides of the same coin, inextricably linked through the fundamental theorem of calculus. Integral calculus and differential calculus empower each other, and a multitude of results and applications to the natural sciences appear. Physics takes off, and the world is now prepared for the Age of Enlightenment, and later, the Industrial Revolution.

The world as we know it would not be even remotely possible without calculus. But let's not forget that what today is calculus was rudimentarily present long before Newton.

"If I have seen further, it is by standing on the shoulders of Giants."
Sir Isaac Newton, in a letter to Robert Hooke, 1675.

Calculus and Education

But bear in mind also that many of the elementary concepts of Calculus were absent until centuries after Newton and Leibniz. Concepts such as function and limit, and the very idea of real number were not formalized until the XIXth Century. The history of calculus has the curious characteristic of being told backwards: the logical order of the concepts of calculus—and almost always the order in which it is taught— is: real number, functions, limit, derivative, and integral, which is almost the reverse of the historical order except for the idea of limit which although formalized in the first half of the XIXth Century was always present in some way or another.

Two hundred years ago there were almost no calculus courses in universities. These were limited to a few military academies and, after the French Revolution, to courses at École Polytechnique and École Normale Supérieure (1794). The usefulness of calculus and its applicability to the most diverse disciplines (engineering, physical science, natural science, and even the social sciences) has led to the fact that the great majority of those who today aspire to a university education must go through some calculus course. And this is fine; for many of those students, calculus will be a tool of daily use, and for others, an unavoidable stepping stone, necessary to advance to other knowledge which requires it.

I believe, however, that calculus has had a negative influence in the education of those who will not need it. In its eagerness to put all students at the doors of calculus, secondary school has often left aside, for lack of time and perhaps lack of usefulness, the more formative parts of elementary mathematics: arithmetic and geometry. Thus, some of the most beautiful pages of mathematics have been replaced by subjects such as absolute value inequalities, and methods for factoring polynomials, justifiable perhaps as a way of preparing students for calculus. In my opinion, more is lost than gained, and calculus should be left to those who will actually use it.

One last comment on calculus courses in the XXIst Century: they deserve to be redesigned from scratch. The typical calculus course today still emphasizes computations. Differentiation and integration of complicated functions can be done with any smartphone. When a smartphone can pass a calculus test, we are not testing (or teaching) correctly. I believe the emphasis must be shifted towards deeper comprehension of the important concepts, and towards interesting applications.

1 The Real Numbers

We begin by noting some properties of the line as a continuum of points, and show the difference between this intuitive line and the line of rational numbers. We then complete the rational line by constructing the real numbers.

The Rational Line

Let us begin with a line:

We all have an intuitive idea of what *the line* is, but to talk about Calculus we need to make precise some of the line's properties and we need to assign numbers to the points of this line. This will not be easy, for the usual sets of numbers (natural, integer, and rational numbers) will not be enough to cover the line that our intuition conceives.

It is customary to represent the numbers according to their order along a line: the natural numbers, which we denote by $\mathbb{N}$:

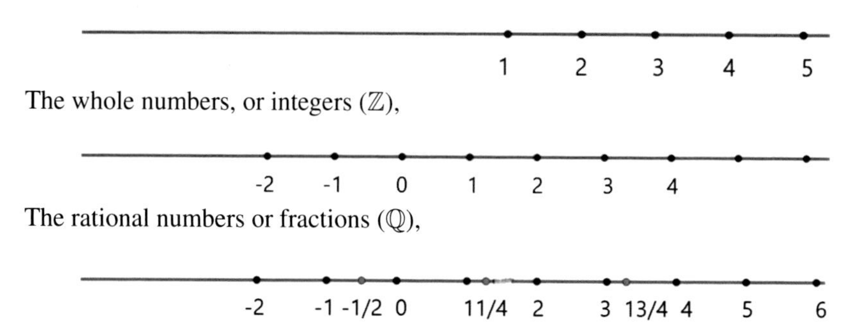

The whole numbers, or integers ($\mathbb{Z}$),

The rational numbers or fractions ($\mathbb{Q}$),

I. Zalduendo, *Calculus off the Beaten Path*, SUMS Readings,
https://doi.org/10.1007/978-3-031-15765-3_1

If we consider the points corresponding to all rational numbers, that is, all those of the form $\frac{k}{n}$ with k integer and n natural, we will have what we call the *rational line* or the *rational points* of the line.

The rational line will not be enough. Many points of what we intuitively recognize as the line are simply not rational points. We will soon see some of these points. But before we do, it will be convenient to consider some properties and some subsets of the line.

Density of $\mathbb{Q}$

The rational line is *dense* in the line: although they are not all the points, the points in the line which correspond to rational numbers, are "all over the place," in the following sense. Consider two points, a and b, in any part of the line, and as close together as you wish:

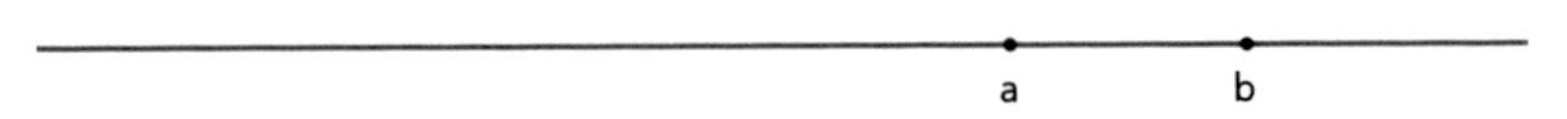

Then there is a point corresponding to a rational number $\frac{k}{n}$ somewhere between a and b. To see this we will find n first and then k. Let's call d the distance between a and b and find a natural number n which is so large that

$$\frac{1}{d} < n.$$

Thus, $\frac{1}{n} < d$. Consider now all rational numbers $\frac{p}{n}$ where p is an integer:

$$\cdots \frac{-2}{n}, \frac{-1}{n}, 0, \frac{1}{n}, \frac{2}{n}, \cdots, \frac{p}{n}, \frac{p+1}{n} \cdots$$

If we set these along our line, we will have a series of points extending infinitely left and right

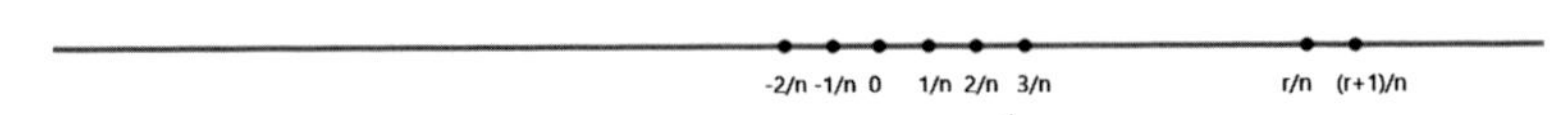

such that the distance between one and the next is $\frac{1}{n}$. But then at least one of them must fall between a and b for the distance between a and b is larger than the distance between two consecutive numbers in our series. We will then have

a k/n b

for some k. It is in this sense that rational points are *dense*.

This means that rational points will serve, for example, to approximate as much as we want, any point a on the line. What do we mean by "as much as we want"? We mean that if given a small positive number, let's call it ε, we will be able to find a rational point whose distance to a is less than ε. And if given a smaller number ε', we will also be able to do it, and for an even smaller ε'', too... Why? Because given any positive (non-zero) ε, consider the points on the line which are at a distance ε from a. There are two (one to each side). Let's call them $a - \varepsilon$ and $a + \varepsilon$. As we have seen, there will be a rational point r between $a - \varepsilon$ and $a + \varepsilon$. The distance from a to r is then smaller than ε:

a-ε r a a+ε

And the point is, we can do this for any ε.

Some Basic Notions

In the preceding argument there have appeared four basic notions that we should point out:

The first is the idea of *order*. Given two different points a and b on the line, one is further left than the other. If a is to the left of b we will write $a < b$ and we will say *a is smaller than b*.

The second notion is *distance*. In the example above we had a smaller than b and we said that the distance between them was d. We will measure distances between points on the line taking as the unit of measure the distance between 0 and 1. For example, the distance from 3/2 to 2 is 1/2 and the distance from 3 to -1 is 4. In general, we define the distance between points a and b as the absolute value of their difference: $|a - b|$, where "absolute value" is defined by

$$|x| = \begin{cases} x, & \text{if } x \geq 0, \\ -x, & \text{if } x < 0. \end{cases}$$

Thus, the distance between a and b is always strictly positive, unless $a = b$, in which case it is zero.

The third notion mentioned above is that of *open interval*: we have drawn the segment

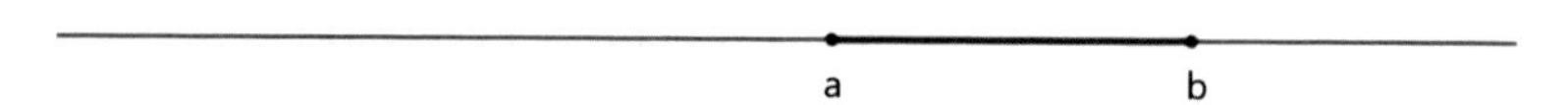

and are interested specifically in all points that are *between* a and b, that is, points x such that $a < x < b$. The set of such points is called the "open interval a, b," and denoted by (a, b). Thus, for example, all points whose distance from a is less than ε form the open interval $(a - \varepsilon, a + \varepsilon)$. Note that neither a nor b belong to the open interval (a, b) for it is not true that $a < a$ nor that $b < b$. Note also that (a, b) contains no "first" element: for each x in (a, b), $a < x$ and hence there is a rational number r such that $a < r < x$. In other words there is something in (a, b) smaller than x. Similarly, (a, b) does not have a largest element. We will also use the "closed interval a, b" which we will denote $[a, b]$ and is made up of all points x such that $a \leq x \leq b$, in other words, we now include a and b.

The fourth notion appearing above is that of *family of neighborhoods* of a point a: we have considered, given any positive ε, the open interval $(a - \varepsilon, a + \varepsilon)$:

Imagine that we now consider smaller and smaller values of ε. We will then have smaller and smaller open intervals... all containing the point a. But no point other than a will be in all of them: indeed, if $b \neq a$, consider $\varepsilon < |b - a|$ and we will have $b \notin (a - \varepsilon, a + \varepsilon)$:

The idea of family of neighborhoods of a will be important when we define limit. There is also another property of the real numbers which we have used: when we choose a natural number n which is "so large that $\frac{1}{d} < n$," we are using the *Archimedean property: given any point on the line, there is a natural number which is larger*. We will prove this in the next chapter, after we have actually constructed the real numbers.

Irrationality of $\sqrt{2}$

We have said above that there are points on the line that do not correspond to any rational number. Let's see one of them: on the segment that joins 0 and 1 we construct a square whose sides measure 1 and then a circle with center at zero and radius equal to the diagonal of the square:

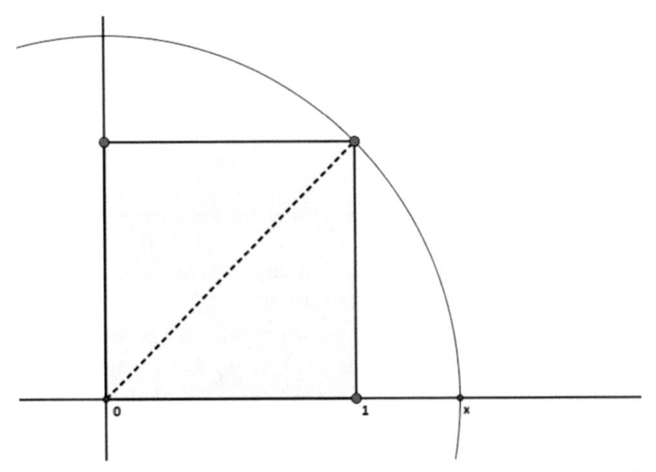

This circle cuts our line at a point whose distance to zero is equal to the length of the diagonal of the square. If we call this length x, we know by the Pythagorean theorem that $x^2 = 1^2 + 1^2 = 2$. This number, which we call "the square root of 2" and denote by $\sqrt{2}$, is not rational. There are several ways to see this. Here's one: suppose $\sqrt{2}$ is rational (we will reach a contradiction) and write

$$\sqrt{2} = \frac{m}{n}$$

with m and n natural numbers. Remember that all fractions may be written in many ways, for example, $\frac{1}{2} = \frac{2}{4} = \frac{3}{6} = \cdots$. Let's agree then to write $\sqrt{2} = \frac{m}{n}$, using the smallest value of n possible. Also since $1 < \sqrt{2} < 2$, i.e., $1 < \frac{m}{n} < 2$, multiplying by n we have: $n < m < 2n$. Now since $\sqrt{2} = \frac{m}{n}$, if we square it we get $2 = \frac{m^2}{n^2}$, in other words $2n^2 = m^2$. Now calculate:

$$\begin{aligned} 2(m-n)^2 &= 2(m^2 - 2mn + n^2) \\ &= 2m^2 - 4mn + 2n^2 \\ &= 4n^2 - 4mn + m^2, \qquad \text{bearing in mind that } 2n^2 = m^2. \\ &= (2n-m)^2. \end{aligned}$$

Then

$$2 = \frac{(2n-m)^2}{(m-n)^2}, \quad \text{that is: } \sqrt{2} = \frac{2n-m}{m-n}.$$

But remember that $n < m < 2n$, so $0 < m - n < n$; this last representation of $\sqrt{2}$ as a fraction has a smaller denominator than n, which was the smallest possible. We have reached a contradiction. This indicates that $\sqrt{2}$ cannot be written as a fraction.

We have therefore found a point on the line that is not a rational point: the point where the circle cuts the line. Is this an anomaly? How common are non-rational points on the line? We will see later that they're much, much more common than rational points. For now, we will just check that non-rational points are also dense on the line. To see this take as before two points $a < b$, and consider the points that correspond to $\frac{a}{\sqrt{2}}$ and to $\frac{b}{\sqrt{2}}$. Since $\frac{a}{\sqrt{2}} < \frac{b}{\sqrt{2}}$, we know there is a rational number r between them: $\frac{a}{\sqrt{2}} < r < \frac{b}{\sqrt{2}}$. Thus

$$a < r\sqrt{2} < b.$$

But this number $r\sqrt{2}$ is not rational. If it were, say $r\sqrt{2} = r'$, we would have $\sqrt{2} = \frac{r'}{r}$, which is also rational, being a quotient of rationals. Thus, in any interval (a, b) there are non-rational points.

From Eudoxus to Dedekind

The rational line—the one that only has rational points—is very deficient if we compare it to the line of our intuition. Its insufficiency resides in the fact that it's full of holes. This we cannot permit for if we intersect with other curves, as, for example, in

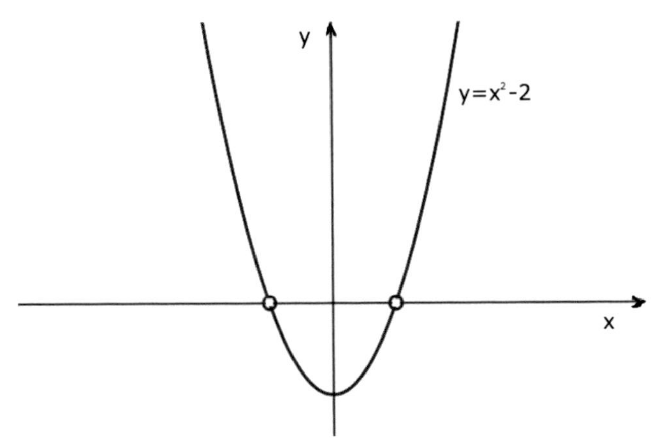

we want to actually have points of intersection!

In the philosophy of the Pythagoreans everything was number. Whole number, or a quotient, rational number. It was thought that any pair of magnitudes of any type, for example, two lengths, were always *commensurable*, meaning they could be measured with the same units, in the following sense: given lengths a and b there would exist a unit r such that $a = mr$ and $b = nr$. But then the quotient of two magnitudes would always be rational

$$\frac{a}{b} = \frac{mr}{nr} = \frac{m}{n}.$$

However, as we have seen, there was the diagonal of a square a and the length of its side b to topple all the philosophy of the Pythagoreans. It is said that it was one of the Pythagoreans, Hippasus of Metapontum who (using the Pythagorean theorem) first showed the irrationality of $\sqrt{2}$. Things seem to have ended very badly for Hippasus who according to different versions: a) was assassinated by the other Pythagoreans, b) committed suicide, or c) drowned at sea after a shipwreck provoked by the gods.

The revolution for Greek mathematics was profound. Arithmetic and numbers lost weight in favor of Geometry. There was, undeniable, the diagonal of the square not corresponding to any number. Magnitudes of different types (lengths, areas, angles) are considered without assigning to them a numerical value. But this is not easy; consider, for example, triangles of the same fixed height

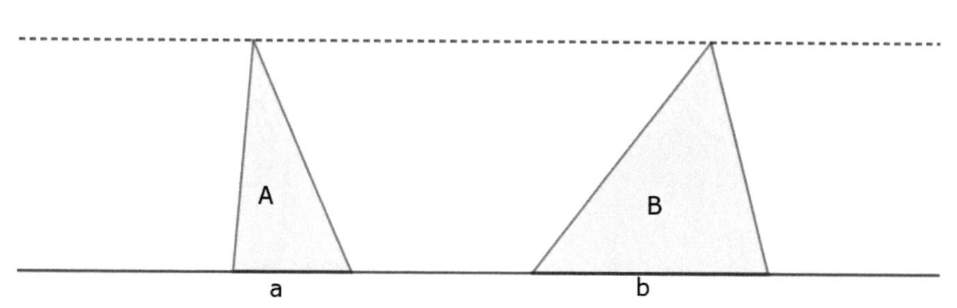

and suppose we want to see that their areas, A and B are proportional to the lengths of their bases, a and b. But the areas are one type of magnitude and the lengths of their bases another. How do we compare $\frac{a}{b}$ with $\frac{A}{B}$ without recurring to the notion of number? Eudoxus of Cnidos invents a way.

Eudoxus (408 AC–355 AC) was a Greek astronomer and mathematician, a student of Archytas of Tarentum. He developed the method of exhaustion proposed by Bryson and Antiphon, and calculated the volumes of pyramids and cones. He introduced the geometric notion of magnitude, and his theory of proportions overcame the deficiencies of the rational numbers by what was one of the first axiomatic presentations of a part of mathematics.

According to his theory of proportions one has $\frac{a}{b} = \frac{A}{B}$ if given whole numbers n and m then:

$$na \text{ is larger, equal to, or less than } mb$$
$$\text{if and only if } nA \text{ is larger, equal to, or less than } mB, \text{ (respectively).}$$

Note that one compares (geometrically) na with mb—magnitudes of the same type—and nA with mB, also magnitudes of the same type and therefore comparable by geometric arguments. Note also that (now thinking in terms of "numbers") $\frac{a}{b}$ and $\frac{A}{B}$ can be irrational, so what Eudoxus' proportionality criterium finally means is

$$\frac{a}{b} \text{ is larger, equal to, or less than } \frac{m}{n}$$
$$\text{if and only if } \frac{A}{B} \text{ is larger, equal to, or less than } \frac{m}{n}, \text{ (respectively).}$$

In other words, $\frac{a}{b} = \frac{A}{B}$ if we would place—on the number line—$\frac{a}{b}$ and $\frac{A}{B}$ on the same side in relation to any rational number $\frac{m}{n}$.

Braunschweig, Germany, 1872. Some 2200 years later, Richard Dedekind finally formalizes the notion of "real number." For Dedekind each real number *is* a "cut": (I, D), that is, a pair of non-empty subsets of the rational numbers with the following properties:

(i) the intersection of I and D is empty but their union is the set of all rational numbers,

(ii) each element of I is less than each element of D, and
(iii) I does not have a maximum element.

For example, the cut (I, D), where

$$I = \{x \in \mathbb{Q} : x < 0 \text{ or } x^2 < 2\}$$

$$\text{and } D = \{x \in \mathbb{Q} : x > 0 \text{ and } x^2 > 2\}$$

is the number $\sqrt{2}$. Two real numbers α and β are equal if

$$\alpha \text{ is larger, equal to, or less than } \frac{m}{n}$$

$$\text{if and only if } \beta \text{ is larger, equal to, or less than } \frac{m}{n}, \text{ (respectively).}$$

In other words, $\alpha = \beta$ if we would place—on the number line—α and β on the same side in relation to any rational number $\frac{m}{n}$. Separated by 2200 years, Eudoxus and Dedekind would have understood each other in five minutes.

In the next section we will make a construction of the real numbers different to Dedekind's.

The Real Line[1]

We have seen that if we pretend to equate points on the line with rational numbers the result is a line full of holes: many points on the line, such as $\sqrt{2}$, simply do not correspond to rational numbers. In this section we will try to explain what a *real number* is. This set of numbers which we will denote $\mathbb{R}$ does seem more like the line (without holes) that we imagine.

2500 years ago, the Greek philosopher Zeno of Elea presented his famous paradox of *Achilles and the turtle*. Achilles is of course the fastest man in the world; the turtle is just any turtle. Zeno argues that if Achilles gives the turtle a head start and they run a race, Achilles will never catch up with the turtle. Zeno argues that when Achilles reaches the point A where the turtle started the race, the turtle will no longer be there, it will be, say, at point B. When Achilles reaches this point the turtle will have advanced to C, and when Achilles reaches C the turtle will be at D... and thus the turtle will *always* be in front of Achilles.

Who would you bet on? The turtle? Let's look then for the error in Zeno's argument. For this, say that the race is along the line, Achilles starts at zero and the turtle starts at the point $\frac{1}{2}$. And suppose also that Achilles is twice as fast as the

[1] This section is reprinted from *Matemática para Iñaki* [13], with kind permission from Fondo de Cultura Económica.

turtle, so that in the time that Achilles runs a certain distance, the turtle runs half that distance. Ready, set, go!

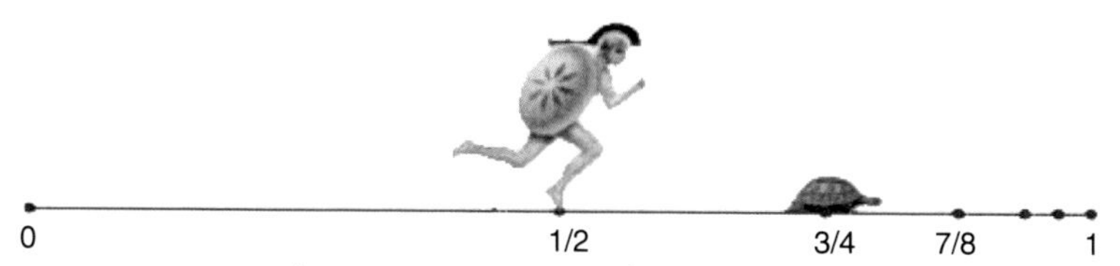

When Achilles reaches $\frac{1}{2}$, the turtle will be at $\frac{3}{4}$, when Achilles reaches $\frac{3}{4}$, the turtle will be at $\frac{7}{8}$, and when Achilles gets to $\frac{7}{8}$, the turtle is at $\frac{15}{16}$, and in general, when Achilles is at the point $\frac{2^n-1}{2^n}$, the turtle is already at $\frac{2^{n+1}-1}{2^{n+1}}$... But won't he catch her? Yes, he will: the time that it takes to go from one point to the next is smaller every time, and although there are infinitely many of these time intervals, they add up to a finite time; the time they take to reach the point 1, which is where Achilles overtakes the turtle.

The error in Zeno's argument is to think that adding infinitely many positive numbers will give infinity. This need not be the case. Consider the lengths just traveled by Achilles at each step: (from 0 to $\frac{1}{2}$, from $\frac{1}{2}$ to $\frac{3}{4}$,...). These lengths are: $\frac{1}{2}, \frac{1}{4}, \frac{1}{8}, \ldots, \frac{1}{2^n}, \ldots$. Let's look at the segment between 0 and 1. If we add all these distances traveled by Achilles, we will get the length of the segment: one.

We will use the following notation: "$\sum$" (the Greek letter *sigma*), means simply to "add." We use it when there are many terms to add and it would be unwieldy to write them all. For example, instead of writing

$$a_1 + a_2 + a_3 + a_4 + a_5 + a_6 + a_7 + a_8 + a_9 + a_{10} + a_{11} + a_{12},$$

we will write

$$\sum_{k=1}^{12} a_k,$$

which reads "the sum with k from 1 to 12, of the numbers a_k." One begins then with k equal to 1 and adds a_1, then for $k = 2$ one adds a_2, for $k = 3$ one adds a_3... until one reaches $k = 12$ and adds the last term, a_{12}. Another example. Instead of writing

$$\frac{1}{2} + \frac{1}{4} + \frac{1}{8} + \cdots + \frac{1}{2^{30}},$$

we will write

$$\sum_{k=1}^{30} \frac{1}{2^k}.$$

We will even add infinitely many terms, which—as we shall see later—sometimes makes sense. Thus we will write

$$\sum_{k=1}^{\infty} a_k.$$

An infinite sum such as this one is called a *series*. I will explain in a special case—geometric series—what this is about. A geometric series is a series of the particular form

$$\sum_{k=0}^{\infty} r^k = 1 + r + r^2 + r^3 + \cdots ,$$

where each term that we are summing is a power of a fixed number r. The series $\frac{1}{2} + \frac{1}{4} + \frac{1}{8} + \cdots + \frac{1}{2^n} + \cdots$ is of this kind, with $r = \frac{1}{2}$ (although we are missing the first term). Sometimes these sums add up to infinity, for example, when r equals 1. Other times they don't. We will say a series is *summable* or *convergent* when its partial sums (up to the nth term)

$$s_n = \sum_{k=0}^{n} r^k$$

converge to some number s (called the sum of the series). Thus, for example,

$$s_n = \frac{1}{2} + \frac{1}{4} + \frac{1}{8} + \cdots + \frac{1}{2^n} \quad \text{converges to } s = 1.$$

It is easy to see that if $0 < r < 1$, the geometric series $\sum_{k=0}^{\infty} r^k$ is summable: note that

$$\begin{aligned}(1-r)s_n &= (1-r)(1 + r + r^2 + \cdots + r^n)\\ &= 1 - r + r - r^2 + r^2 - r^3 + \cdots + r^n - r^{n+1}\\ &= 1 - r^{n+1}.\end{aligned}$$

Then

$$s_n = \frac{1 - r^{n+1}}{1 - r}.$$

But as $0 < r < 1$, r^{n+1} tends to zero as n grows. Therefore the partial sums s_n converge to $\frac{1}{1-r}$. We will write

$$\sum_{k=0}^{\infty} r^k = \frac{1}{1-r}.$$

In the case where $r = \frac{1}{2}$ this sums 2. And if instead of summing from $k = 0$ we sum from $k = 1$, it sums 1. In other words

$$\frac{1}{2} + \frac{1}{4} + \frac{1}{8} + \frac{1}{16} + \cdots = 1.$$

Note that not only the terms in the sum get smaller and smaller but also the "tails" $\sum_{k>n} \frac{1}{2^k}$ get smaller. In fact the tail $\sum_{k>n} \frac{1}{2^k}$, in which we sum from $\frac{1}{2^{n+1}}$, adds $\frac{1}{2^n}$: indeed,

$$\begin{aligned}\sum_{k>n} \frac{1}{2^k} &= \frac{1}{2^{n+1}} + \frac{1}{2^{n+2}} + \frac{1}{2^{n+3}} + \cdots \\ &= \frac{1}{2^n 2} + \frac{1}{2^n 4} + \frac{1}{2^n 8} + \cdots \\ &= \frac{1}{2^n}\left(\frac{1}{2} + \frac{1}{4} + \frac{1}{8} + \cdots\right) \\ &= \frac{1}{2^n},\end{aligned}$$

for the sum in parenthesis adds to 1.

Dyadic Series—A Construction of $\mathbb{R}$

Consider the geometric series with $r = \frac{1}{2}$, but let's say that now some of the terms we add and others we don't. For example:

$$\frac{1}{2} + \frac{0}{4} + \frac{1}{8} + \frac{0}{16} + \cdots.$$

The sum will no longer be 1, but less. In fact by choosing to add or not add each term, in other words by setting each d_k equal to 1 or 0 in the following series

$$\sum_{k=1}^{\infty} \frac{d_k}{2^k} = \frac{d_1}{2} + \frac{d_2}{4} + \frac{d_3}{8} + \frac{d_4}{16} + \cdots,$$

we will be able to reach any point between 0 and 1. For us the real numbers between 0 and 1 will *be* these dyadic series. For example, suppose we want to obtain the point marked on the line:

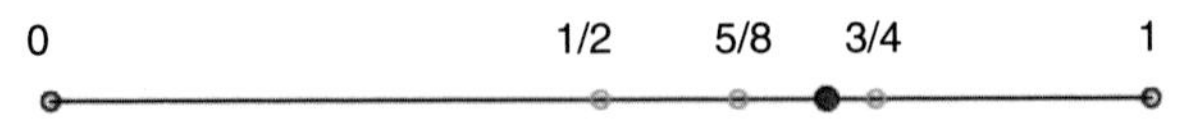

Since it is larger than $\frac{1}{2}$, we will set $d_1 = 1$; since it's less than $\frac{1}{2}+\frac{1}{4}$, we put $d_2 = 0$; but as it is larger than $\frac{1}{2}+\frac{0}{4}+\frac{1}{8}$, we write $d_3 = 1\ldots$ and thus we may approximate the point as much as we desire. The series

$$\sum_{k=0}^{\infty} \frac{d_k}{2^k}$$

will correspond to the point we want. The following argument will show how $d_1, d_2, d_3, \ldots$ must be chosen in order for the sum to be a given rational number, $\frac{m}{n}$, between 0 and 1. More importantly, the argument will show that for the sum to be a rational number, at some point the digits d_k will begin to repeat themselves. Since $\frac{m}{n} < 1$, $2m < 2n$. Thus if we divide $2m$ by n, using the division algorithm we have

$$2m = d_1 n + r_1 \quad \text{with } d_1 = 0 \text{ or } 1, \text{ and } 0 \leq r_1 < n.$$

Here d_1 is the quotient, and r_1 the remainder. Since $r_1 < n$, we have $2r_1 < 2n$. If we now divide $2r_1$ by n (and analogously for $r_2, r_3, \ldots$) we have

$$\begin{aligned}
2r_1 &= d_2 n + r_2 \quad \text{with } d_2 = 0 \text{ or } 1, \text{ and } 0 \leq r_2 < n, \\
2r_2 &= d_3 n + r_3 \quad \text{with } d_3 = 0 \text{ or } 1, \text{ and } 0 \leq r_3 < n, \\
&\vdots \\
2r_{k-1} &= d_k n + r_k \\
&\vdots
\end{aligned}$$

We then have,

$$\begin{aligned}
\frac{m}{n} &= \frac{d_1}{2} + \frac{r_1}{2n} \quad \text{(dividing } 2m \text{ above by } 2n) \\
&= \frac{d_1}{2} + \frac{d_2}{4} + \frac{r_2}{4n} \quad \text{(dividing } 2r_1 \text{ by } 4n) \\
&= \frac{d_1}{2} + \frac{d_2}{4} + \frac{d_3}{8} + \frac{r_3}{8n} \quad \text{(dividing } 2r_2 \text{ by } 8n) \\
&\vdots \\
&= \frac{d_1}{2} + \frac{d_2}{4} + \frac{d_3}{8} + \cdots + \frac{d_k}{2^k} + \frac{r_k}{2^k n} \quad \text{(dividing } 2r_{k-1} \text{ by } 2^k n.)
\end{aligned}$$

Note that $\frac{r_k}{2^k n} < \frac{1}{2^k}$, so in summing more and more terms we have

$$\frac{m}{n} = \sum_{k=1}^{\infty} \frac{d_k}{2^k}.$$

Now, note that each d_k depends only on the remainder r_{k-1} in the previous step. For this reason the digits d_k at some moment start to repeat themselves. Why? Because the remainders must repeat themselves: there are only n possible remainders: $0, 1, 2, \ldots, n-1$. And as soon as one remainder, r, is repeated, all the steps from that previous appearance of r will be repeated. So each rational number between 0 and 1 corresponds to a sum

$$\sum_{k=1}^{\infty} \frac{d_k}{2^k},$$

where the sequence $d_1, d_2, d_3, \ldots$ finally repeats itself. But there are of course many other sequences $d_1, d_2, d_3, \ldots$ that never repeat themselves. For example:

$$1, 0, 1, 1, 0, 1, 1, 1, 0, 1, 1, 1, 1, 0, 1, 1, 1, 1, 1, 0, \ldots$$

In some sense, almost all sequences are non-repeating sequences, and they correspond to numbers

$$x = \sum_{k=1}^{\infty} \frac{d_k}{2^k}$$

that are not rational. Considering the set of all such dyadic series we fix all the "holes" left by the rational numbers between 0 and 1, and if to these we add integers k, we fill in all the holes between k and $k+1$. These numbers that we are filling in are called *irrational* (such as $\sqrt{2}$). The rational numbers together with the irrational numbers form the set of all real numbers. Now at last we have the line with no holes in it: the *real line* $\mathbb{R}$.

The Scarcity of $\mathbb{Q}$

The following heuristic argument will give us an idea of how small the set of rational numbers is within the real line. We will construct numbers in the interval $(0, 1)$ using the following probabilistic method: we throw a coin infinitely many times and write down 1 each time we obtain heads, and 0 when we obtain tails. For example:

$$0, 1, 1, 0, 1, 0, 1, 0, 0, 0, 1, 1, 0, 1, 1, 1, 0, 0, 1, \ldots$$

and now we use this sequence of zeros and ones to obtain the number

$$\sum_{k=1}^{\infty} \frac{d_k}{2^k}, \text{ where } d_k = \begin{cases} 1, & \text{if on the } k\text{-th throw we obtained heads,} \\ 0, & \text{if on the } k\text{-th throw we obtained tails.} \end{cases}$$

Thus every number in the interval $(0, 1)$ corresponds to the result of an experiment which consisted in throwing a coin infinitely many times, and each subset corresponds to a set of possible outcomes of that experiment. For example, the open interval $(0, \frac{1}{2})$ is formed by points obtained in experiments where $d_1 = 0$: the first throw was tails; the rest, whatever. What is the probability of this occurrence? It is $\frac{1}{2}$. The interval $(\frac{5}{8}, \frac{6}{8})$ is formed by points obtained in experiments where $d_1 = 1, d_2 = 0$, and $d_3 = 1$: we have obtained heads, tails, heads; and then anything else. What is the probability of this occurring? It is $\frac{1}{2} \times \frac{1}{2} \times \frac{1}{2} = \frac{1}{8}$. In general, the measure of a set $A \subset (0, 1)$ is equal to the probability of producing one of its points with our experiment of throwing a coin infinitely many times.

Say now that $A = \{r \in \mathbb{Q} : 0 < r < 1\}$, the rational numbers of the interval $(0, 1)$. We have just seen that each of them is obtained with a sequence

$$d_1, d_2, d_3, d_4, \ldots$$

which is finally periodic. You throw a coin infinitely many times; from a certain moment onward, for example, after the 37th throw, the results start to repeat themselves. . . and they repeat. . . and repeat. . . and repeat. . . infinitely. What is the chance of this happening? Zero. That's the measure of the set of rational numbers: zero.

The Completeness of $\mathbb{R}$

"Completeness" is what marks the difference between the real line and the rational line. We will show here that the real line $\mathbb{R}$, which we have constructed with the dyadic series, is "complete."

Given a subset A of $\mathbb{R}$, we say that the number c is a *bound* or, more precisely, an *upper bound* of A if c is to the right of all of set A, in other words if $a \leq c$ for all elements $a \in A$. Suppose we have a non-empty set A such that 1 is a bound, but 0 is not. That is, there's some number $a \in A$ such that $0 < a \leq 1$, but nothing to the right of 1. We will construct a number s which is the least of all upper bounds. We will write it as

$$s = \sum_{k=1}^{\infty} \frac{d_k}{2^k},$$

by choosing very carefully the d_k's. We will obtain that s is a bound for A, but none of the partial sums

$$s_n = \sum_{k=1}^{n} \frac{d_k}{2^k}$$

will be a bound. Choose the d_k's in the following way:

$$d_1 = \begin{cases} 0, & \text{if } \frac{1}{2} \text{ is a bound,} \\ 1, & \text{if not.} \end{cases}$$

Thus, $s_1 = \frac{d_1}{2}$ is not a bound for A...

$$d_2 = \begin{cases} 0, & \text{if } s_1 + \frac{1}{4} \text{ is a bound,} \\ 1, & \text{if not.} \end{cases}$$

Thus, $s_2 = s_1 + \frac{d_2}{4}$ is not a bound... In the n-th step,

$$d_n = \begin{cases} 0, & \text{if } s_{n-1} + \frac{1}{2^n} \text{ is a bound,} \\ 1, & \text{if not.} \end{cases}$$

Thus, $s_n = s_{n-1} + \frac{d_n}{2^n}$ is not a bound...

Note that the partial sums s_n are not bounds, and they converge, increasing, to s. We will see that s is an upper bound. If it were not, we would have $s < a$ for some element a in the set A. In that case, if n is so large that

$$\frac{1}{2^n} < a - s,$$

we will have

$$s_{n-1} + \frac{1}{2^n} \leq s + \frac{1}{2^n} < a.$$

Now, since $s_{n-1} + \frac{1}{2^n}$ is not a bound, we will have chosen $d_n = 1$. This for each sufficiently large n. Then the sequence $\ldots d_k \ldots$ is finally constant; after a certain point, a sequence of ones. Consider the last zero: we have m such that $d_m = 0$, and after that, all ones. Since $d_m = 0$ we know that

$$s_{m-1} + \frac{1}{2^m} \quad \text{is a bound.}$$

But

$$s = s_{m-1} + \frac{0}{2^m} + \frac{1}{2^{m+1}} + \frac{1}{2^{m+2}} + \frac{1}{2^{m+3}} + \cdots$$

$$=s_{m-1} + \frac{1}{2^m}\left(\frac{1}{2} + \frac{1}{4} + \frac{1}{8} + \cdots\right)$$

$$=s_{m-1} + \frac{1}{2^m}, \qquad \text{which is a bound.}$$

Finally, s is the least upper bound, for if $c < s$ were a bound, for sufficiently large n, $c < s_n \leq s$, and s_n would be a bound, but we know it is not.

The number s we have constructed is the least upper bound of the set A, also called *supremum* of A. In other words there is a number s (which may or may not belong to the set A) such that s is a bound of A, and if $b < s$, b is not a bound, that is, there is an element $a \in A$ such that $b < a \leq s$. The difference between the real line and the rational line is that on the real line every non-empty subset A that has an upper bound will also have a least upper bound or supremum. On the rational line this is not true: for example, the set

$$A = \{r \in \mathbb{Q} : r^2 < 2\}$$

has a bound, but no least upper bound. What happens is that this "least upper bound" would be $s = \sqrt{2}$, but that's where $\mathbb{Q}$ has a hole. This property is usually called the "Completeness Axiom." It will be a fundamental tool in several very basic theorems about real numbers, for example, the theorem on bounded increasing sequences and Bolzano's Theorem. We have then, the

Completeness Axiom *Every non-empty subset of the real line having an upper bound has a supremum (a least upper bound).*

Cardinality

We take a detour here for a couple of comments about cardinality. How many elements does a set have? How many eggs are still in the refrigerator? How many players has the other team brought? We answer these questions by counting: we point at each player and say "one, two, three,..." until there are no more. What we do when we count is establish a correspondence between the objects that we wish to count and some set of natural numbers, $\{1, 2, 3, \ldots, 14\}$: they have 14 players. In order to count correctly, the correspondence we establish must be one to one and onto. Each object corresponds to a different number, each number to a different object. We are defining what we call a bijective function between the objects that we wish to count and the set of the first n natural numbers.

Two sets have the same number of elements—the same cardinality—if we can establish between those two sets a bijective correspondence. Here's an example between sets with infinitely many elements:

Example The set of all integers, $\mathbb{Z}$, and the set of all natural numbers, $\mathbb{N}$, have the same cardinality. We can establish the following correspondence between them

$$f : \mathbb{Z} \longrightarrow \mathbb{N}, \quad \text{defined like this:} \quad f(k) = \begin{cases} 2k+1, & \text{if } k \geq 0 \\ -2k, & \text{if } k < 0. \end{cases}$$

In a drawing:

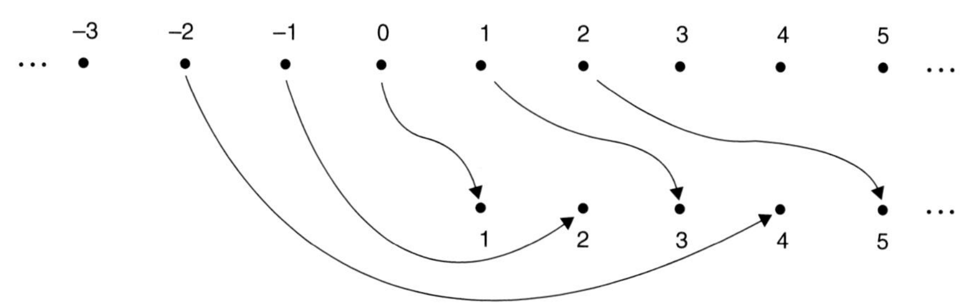

It may surprise you that $\mathbb{N}$, which is a subset of $\mathbb{Z}$, has as many elements as $\mathbb{Z}$. Among finite sets this does not happen, but you'll see that when the sets are infinite it is not uncommon. Also the correspondences

$$n \mapsto 2n \mapsto 2n - 1$$

between

$$\mathbb{N} \longrightarrow \{\text{even natural numbers}\} \longrightarrow \{\text{odd natural numbers}\}$$

are bijective, thus the three sets have the same cardinality, although we have

$$\mathbb{N} = \{\text{even natural numbers}\} \cup \{\text{odd natural numbers}\}.$$

However, not all infinite sets have the same cardinality. Some infinite sets are larger than others, as we will see below.

When there is a bijective correspondence between $\mathbb{N}$ and a set A we will say that A is *countable*. Its elements can be numbered with the natural numbers: if $f : \mathbb{N} \longrightarrow A$ is bijective, $A = \{f(1), f(2), f(3), \ldots\}$. If a set is countable, we usually write

$$A = \{a_1, a_2, a_3, \ldots\}.$$

We have seen that $\mathbb{Q}$ is, in the sense of measure, a small subset of $\mathbb{R}$: its measure is zero. We will now see that it is also small in the sense of cardinality. Although $\mathbb{Q}$ and $\mathbb{R}$ are both infinite sets, $\mathbb{Q}$ is countable but $\mathbb{R}$ is not. The infinity of the real numbers is larger than that of the rational numbers. As it turns out, there are different categories of infinity.

Let's first see that $\mathbb{Q}$ is countable. Consider the following table containing all positive rationals. The numbers on top determine the numerators, those on the left, the denominators.

	1	2	3	4	5	6	7	8	9	
1	1/1	2/1	3/1	4/1	5/1	6/1	7/1	....	....	....
2	1/2	2/2	3/2	4/2	5/2	6/2	....	....	....	....
3	1/3	2/3	3/3	4/3	5/3	6/3	....	....	....	....
4	1/4	2/4	3/4	4/4	5/4	....	....	....	....	....
5	1/5	2/5	3/5	4/5	5/5	....	....	....	....	....
6	1/6	2/6	3/6	4/6	....	....	....	....	....	....
7	1/7	2/7	3/7	....	....	....	....	....	....	....

Of course, each appears several times, for example, $\frac{1}{2} = \frac{2}{4} = \frac{3}{6} = \cdots$. But we may visit all positive rational numbers by following the arrows in the graph and assign an odd natural number to each new fraction that we find (omitting the rational numbers that we have already numbered). We then number the negative rationals with the even natural numbers: $r_{2n} = -r_{2n-1}$. Thus we may write

$$\mathbb{Q} = \{r_1, r_2, r_3, \dots\}.$$

But the real numbers are not countable. In fact the open interval $(0, 1)$ of real numbers between 0 and 1 is uncountable. Let's see why. Suppose, towards a contradiction, that $(0, 1)$ were countable, and write

$$(0, 1) = \{x_1, x_2, x_3, \dots\}.$$

Now write the decimal expression of each number:

$$x_1 = 0, d_{11}d_{12}d_{13}\dots$$

$$x_2 = 0, d_{21}d_{22}d_{23}\dots$$
$$x_3 = 0, d_{31}d_{32}d_{33}\dots$$
$$\vdots$$
$$x_n = 0, d_{n1}d_{n2}d_{n3}\dots d_{nn}\dots$$
$$\vdots$$

Choose for each k, a digit d_k (say 3, or 7) different from d_{kk}, and define the number $y = \sum_{k=1}^{\infty} \frac{d_k}{10^k}$. The decimal expression of y is $y = 0, d_1d_2d_3\dots$. Clearly $y \in (0, 1)$. But it is not x_1, because $d_1 \neq d_{11}$, it's not x_2, for $d_2 \neq d_{22}$, not x_3, for $d_3 \neq d_{33}$, …not x_n, because $d_n \neq d_{nn}$…it is none of the x_n's. This is a contradiction, for we had numbered all the elements in the set $(0, 1)$. Thus $(0, 1)$ is not countable.

Exercises

1 Prove that

(a) $|x + y| \leq |x| + |y|$.
(b) $|ax| = |a||x|$.
(c) $|x| \geq 0$ and $= 0$ if and only if $x = 0$.

2 Prove that $||x| - |y|| \leq |x - y|$.

3 Given a natural number n, prove that if n^2 is even, then so is n.

4 Another proof of the irrationality of $\sqrt{2}$: start by supposing that $\sqrt{2} = \frac{m}{n}$, where m and n are not both even…Use Exercise 3 to arrive at a contradiction.

5 Prove that $\sqrt{3}$ is irrational.

6 Prove that $s = \sup A$ if and only if s is an upper bound of A and for each $\varepsilon > 0$ we may find $a_\varepsilon \in A$ such that $s - \varepsilon < a_\varepsilon \leq s$.

7 In analogy to the definition of $\sup A$ define $i = \inf A$ (*infimum* of A) setting

(a) i is a lower bound of A, and
(b) if $i < b$, b is not a lower bound of A.

If B is a (upper and lower) bounded set, and $A \subset B$, determine (and prove) the order of

$$\sup A \qquad \inf A \qquad \sup B \qquad \inf B$$

8 If $-A = \{-x : x \in A\}$, determine (and prove) the relationships between

$$\sup(-A) \qquad -\sup A \qquad \inf(-A) \qquad -\inf A$$

Sequences and Series

2

In this chapter we introduce the notion of limit of a sequence of points and study some topological properties of closed intervals of the real line. We then consider the convergence of series, including absolute and unconditional convergence.

Sequences

We will need to be precise regarding the limit of a sequence. Actually we have already seen a few: we have said that the numbers $\frac{1}{n}$ tend to zero and also that the partial sums of the geometric series

$$s_n = \sum_{k=0}^{n} r^k$$

tend to $\frac{1}{1-r}$ when $0 < r < 1$.

In the first decades of the XIXth Century it became clear that the notion of limit, which had been rudimentary present in mathematics since the times of the ancient Greeks, required formalization and precise definitions. Bolzano and Cauchy finally provided mathematics with the notion of limit.

Augustin-Louis Cauchy (1789–1857) was born in Paris at the time of the French Revolution. His family escaped to Arcueil, where Augustin spent his boyhood. Back in Paris he studied at the École Polytechnique, and later at the École des Ponts et Chaussées. Although he worked several years as an engineer, he soon became more strongly attracted to Physics and Mathematics. A prolific author, he made important contributions to the foundations of calculus and to complex analysis.

I. Zalduendo, *Calculus off the Beaten Path*, SUMS Readings,
https://doi.org/10.1007/978-3-031-15765-3_2

Limits of Sequences

In general, given a sequence $a_1, a_2, a_3, \ldots, a_n, \ldots$ of real numbers (that is, a function assigning to each natural number n a real number a_n), we will say that a_n *tends* to a, or that a_n *converges* to a (we will write "$a_n \longrightarrow a$"), or that a is the *limit* of the sequence a_n ($\lim_n a_n = a$) if:

Given *any* neighborhood of a, the sequence a_n *finally* stays in that neighborhood.

As you can see, I have emphasized two words which are key in this definition: "any" and "finally." Recall that the neighborhoods of a are the open intervals

$$(a - \varepsilon, a + \varepsilon), \quad \text{with } \varepsilon > 0.$$

When we say that $a_n \longrightarrow a$ it means that—given any $\varepsilon_0 > 0$—we will "finally" have $a_n \in (a - \varepsilon_0, a + \varepsilon_0)$ after a certain value of n; say for all $n \geq n_0$. If given a smaller value of ε, say $\varepsilon_1 < \varepsilon_0$, the same will happen after a certain (perhaps larger) value of n: $a_n \in (a - \varepsilon_1, a + \varepsilon_1)$ for all $n \geq n_1$. And if $\varepsilon_2 < \varepsilon_1$, we will have $a_n \in (a - \varepsilon_2, a + \varepsilon_2)$ for all $n \geq n_2$, etc. Thus we may assure that the distance between a_n and a may be made as small as required by simply taking n sufficiently large:

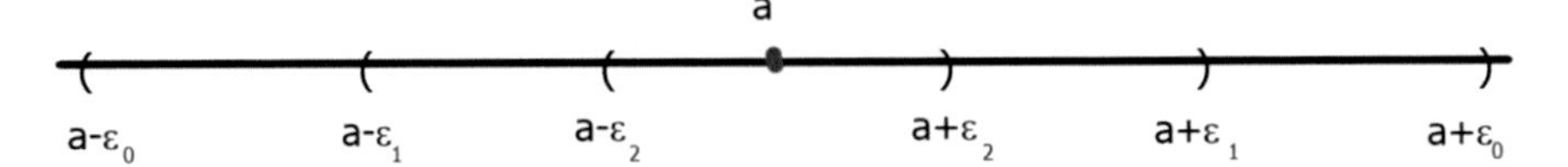

$$|a_n - a| < \varepsilon \text{ for all } n \geq n_\varepsilon.$$

Let's see this in the examples we've already mentioned:

Example $\frac{1}{n} \longrightarrow 0$. Given $0 < \varepsilon < 1$, let's write the dyadic expression of ε:

$$\varepsilon = \sum_{k=1}^{\infty} \frac{d_k}{2^k},$$

and consider the smallest index k_0 such that d_{k_0} is one. Then,

$$\varepsilon = \frac{1}{2^{k_0}} + \sum_{k=k_0+1}^{\infty} \frac{d_k}{2^k} \geq \frac{1}{2^{k_0}} > \frac{1}{2^{k_0} + 1},$$

and for all $n > 2^{k_0}$ we will also have $-\varepsilon < \frac{1}{n} < \varepsilon$.

Note that with this we have also proved the *Archimedean property*: the set of natural numbers $\mathbb{N}$ is not bounded above within the set of real numbers, for if $1 < a \in \mathbb{R}$, for some n we will have $\frac{1}{n} < \varepsilon = \frac{1}{a}$, that is: $a < n$.

Example $s_n = \sum_{k=0}^{n} r^k \longrightarrow \frac{1}{1-r}$. Given $\varepsilon > 0$,

$$\frac{1}{1-r} - \varepsilon < s_n < \frac{1}{1-r} + \varepsilon$$

if (recall that $s_n = \frac{1-r^{n+1}}{1-r} = \frac{1}{1-r} - \frac{r^{n+1}}{1-r}$)

$$\frac{1}{1-r} - \varepsilon < \frac{1}{1-r} - \frac{r^{n+1}}{1-r} < \frac{1}{1-r} + \varepsilon$$

that is $\frac{r^{n+1}}{1-r} < \varepsilon$, in other words when $r^{n+1} < \varepsilon(1-r)$, which happens if n is sufficiently large, for r is smaller than 1. Here, the required values of n are $n > \frac{\log(\varepsilon(1-r))}{\log r} - 1$... In general, we will not care for the exact value of n necessary, but simply notice that the inequality will hold for *sufficiently large* values of n.

Of course not all sequences converge. Two examples:

(i) $a_n = n$: becomes larger and larger but does not approach any real number.
(ii) $a_n = (-1)^n$: is -1 for odd n and 1 for even n; this sequence jumps from 1 to -1 and back from -1 to 1 repeatedly... it cannot remain in any interval (a, b) of length less than 2.

We now prove the following fact:

Proposition *Given two converging sequences, $a_n \longrightarrow a$ and $c_n \longrightarrow c$; such that for all n we have $a_n \leq c_n$, then $a \leq c$.*

Let's see why: if the conclusion were not to hold, we would have $c < a$. Consider then $\varepsilon > 0$ so small that the open intervals $(c - \varepsilon, c + \varepsilon)$ and $(a - \varepsilon, a + \varepsilon)$ do not intersect ($\varepsilon < \frac{a-c}{2}$ will do):

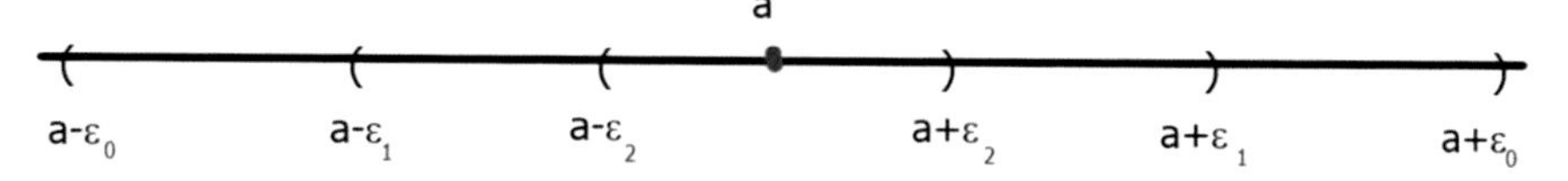

Since $c_n \longrightarrow c$ and $a_n \longrightarrow a$, for sufficiently large n, c_n will be in the neighborhood of c and a_n in the neighborhood of a. We will then have $c_n < a_n$, contradicting the fact that $a_n \leq c_n$ for all n. □

Among the exercises to this chapter you will find other properties that will also be useful. For example:

If $a_n \longrightarrow a$ and $b_n \longrightarrow b$,

(i) $a_n + b_n \longrightarrow a + b$,
(ii) $a_n b_n \longrightarrow ab$,
(iii) if $a = b$ and $a_n \leq c_n \leq b_n$, then $c_n \longrightarrow a = b$.

We now come to an important Theorem:

Increasing Bounded Sequence Theorem *Every increasing and bounded (from above) sequence converges.*

Let's see why. If a_n is an increasing sequence ($a_n \leq a_{n+1}$) and is bounded ($a_n \leq c$ for all n), consider the set

$$A = \{a_n : n \in \mathbb{N}\}.$$

A is a non-empty bounded set (each element of A is $\leq c$). Thus, by the completeness axiom, the supremum of A exists. Call it s. We will check that then $a_n \longrightarrow s$. To this end, take $\varepsilon > 0$; we wish to see that

$$s - \varepsilon < a_n < s + \varepsilon$$

for all sufficiently large n. The right-hand inequality always holds, for $a_n \leq s$ for all n. The left inequality: as s is the supremum of A, nothing smaller than s can be an upper bound of A. Hence, $s - \varepsilon$ is not, so for some n_0, $s - \varepsilon < a_{n_0}$. But for all n larger than n_0 also, for the sequence a_n is increasing. Therefore, for all $n \geq n_0$ we have

$$s - \varepsilon < a_n < s + \varepsilon.$$

□

Analogously, every sequence that is decreasing and bounded (below) converges. These results do not hold in the rational line (which is not complete). For example, if the decimal expression of $\sqrt{2}$ is

$$\sqrt{2} = 1, d_1 d_2 d_3 \ldots d_n d_{n+1} \ldots$$

the sequence (a_n) with $a_n = 1, d_1 d_2 d_3 \ldots d_n$ is increasing and bounded but does not converge in $\mathbb{Q}$.

Cantor's Nested Intervals Theorem

Recall that we have called the set

$$[a, b] = \{x \in \mathbb{R} : a \leq x \leq b\},$$

the "closed interval a b."

Cantor's Nested Intervals Theorem *If $I_n = [a_n, b_n]$ is a sequence of decreasing closed intervals (that is, $I_n \supset I_{n+1}$ for all n), then the intersection of all these intervals is non-empty:*

$$\bigcap_{n \in \mathbb{N}} I_n \neq \emptyset.$$

If, moreover, the lengths of the intervals $(b_n - a_n)$ tend to zero, the intersection contains only one point.

Let's see why: the inclusions that we have between our intervals imply that $a_n \leq a_{n+1} \leq b_{n+1} \leq b_n$ for all n. Thus, the sequence (a_n) is increasing and bounded above (by b_1), while the sequence (b_n) is decreasing and bounded below (by a_1). Thus they both converge. Say $a_n \to a$ and $b_n \to b$. Now since for every n, $a_n \leq b_n$, we have $a \leq b$, and

$$[a, b] \subset \bigcap_{n \in \mathbb{N}} I_n,$$

since for every n, $a_n \leq a \leq b \leq b_n$. Also, if the lengths of the intervals tend to zero, we will have

$$b - a \leq b_n - a_n \to 0,$$

thus $a = b$, and the intersection contains only one point. □

Note that Cantor's nested intervals theorem would not hold without completeness. Using, as above, the decimal expression for $\sqrt{2}$, and setting

$$I_n = \left[a_n, a_n + \frac{1}{10^n}\right],$$

we see that it does not hold in $\mathbb{Q}$: in this case we have (using the notation in the theorem), $a = \sqrt{2} = b$.

Subsequences

We will need the notion of *subsequence*. We have said that a sequence (a_n) is a function assigning a real number a_n to each natural number n:

$$1 \mapsto a_1$$

$$2 \mapsto a_2$$
$$3 \mapsto a_3$$
$$\vdots$$

For example, the sequence $a_n = (-1)^n$ is the correspondence:

$$\begin{aligned} 1 &\mapsto -1 \\ 2 &\mapsto 1 \\ 3 &\mapsto -1 \\ 4 &\mapsto 1 \\ 5 &\mapsto -1 \\ &\vdots \end{aligned}$$

A subsequence of the sequence (a_n) is a correspondence which consists in first taking a strictly increasing sequence of indices ($n_1 < n_2 < n_3 < \cdots$), and then the original correspondence $n \mapsto a_n$:

$$\begin{aligned} 1 &\mapsto n_1 \mapsto a_{n_1} \\ 2 &\mapsto n_2 \mapsto a_{n_2} \\ 3 &\mapsto n_3 \mapsto a_{n_3} \\ &\vdots \end{aligned}$$

For example, the sequence 1, 1, 1, 1... is a subsequence of $a_n = (-1)^n$, considering

$$\begin{aligned} 1 &\mapsto 2 \mapsto (-1)^2 \\ 2 &\mapsto 4 \mapsto (-1)^4 \\ &\vdots \\ k &\mapsto 2k \mapsto (-1)^{2k} \\ &\vdots \end{aligned}$$

and the sequence $1, \frac{1}{2}, \frac{1}{4}, \frac{1}{8}, \frac{1}{16}, \ldots$ is a subsequence of $a_n = \frac{1}{n}$, for we may write

$$k \mapsto 2^k \mapsto \frac{1}{2^k}.$$

We will use the notation (a_{n_k}) to indicate a subsequence of (a_n):

$$k \mapsto n_k \mapsto a_{n_k},$$

(where $n_1 < n_2 < n_3 \ldots$).

Note that a non-convergent sequence may have subsequences which do converge. The following result shows that if the original sequence is contained in a closed interval, this always happens.

Theorem *If* $(a_n) \subset [a, b]$, (a_n) *has a subsequence converging to a point of* $[a, b]$.

Let's see why. If the sequence (a_n) has only finitely many points in its image, that is, if

$$\{a_n : n \in \mathbb{N}\} = \{c, d, e, f, \ldots, z\},$$

doubtless one of them will be repeated infinitely many times. Say, for example, that $d = a_{n_k}$ for infinitely many n_k. Then $d, d, d, \ldots$ is a convergent subsequence of (a_n). If, on the contrary, the image of the sequence ($\{a_n : n \in \mathbb{N}\}$) has infinitely many points, we proceed as follows. Split the interval $[a, b]$ into two halves: two closed intervals of length $\frac{b-a}{2}$. One of the two, which we will call I_1, contains infinitely many points of the sequence (a_n). Now split I_1 into half; we have two closed intervals of length $\frac{b-a}{4}$. And, as before, one of them has infinitely many points of the sequence. Call it I_2. Split I_2 into two halves, one of which contains infinitely many points of (a_n)... By continuing this process we obtain a sequence of closed intervals (I_n) which is decreasing: $I_n \supset I_{n+1}$. Also, the lengths of the I_n are $\frac{b-a}{2^n}$, which tend to zero. Thus by Cantor's nested intervals theorem, there will be just one point in their intersection:

$$\bigcap_{n \in \mathbb{N}} I_n = \{x\}.$$

But we have chosen the I_n's in such a way that each contains infinitely many of the points of the sequence (a_n). We may then take a subsequence (a_{n_k}) such that

$$a_{n_k} \in I_k \quad \text{for each } k.$$

This subsequence converges to x. □

Series

We have seen—in the case of the geometric series $\sum_{k=0}^{\infty} r^k$, with $0 < r < 1$—that sometimes summing infinitely many numbers makes sense. Other times it doesn't. We may ask ourselves, for example, if we can assign a "sum" S to

$$\sum_{k=0}^{\infty}(-1)^k = 1 - 1 + 1 - 1 + 1 - 1 + 1 - \cdots = S.$$

On the one hand, we would expect that $S = (1-1) + (1-1) + \cdots = 0$, although also

$$\begin{aligned} -S &= -1 + 1 - 1 + 1 - 1 + 1 - 1 + \cdots \\ &= -1 + (1 - 1 + 1 - 1 + 1 - 1 \cdots) \\ &= -1 + S, \end{aligned}$$

from where $S = \frac{1}{2} \ldots$

This means that we need to be more precise about what we mean by the sum of a series $\sum_{k=1}^{\infty} a_k$. We will adopt the definition that we used in Chap. 1 to sum the geometric series. We say that a series *converges* and that its sum is s, if the sequence given by its partial sums

$$s_n = \sum_{k=1}^{n} a_k$$

tends to the number s. Thus the geometric series $\sum_{k=0}^{\infty} r^k$ converges if $0 < r < 1$, and its sum is $\frac{1}{1-r}$. The series $\sum_{k=0}^{\infty}(-1)^k$ has partial sums

$$\sum_{k=0}^{n}(-1)^k = s_n = \begin{cases} 1, & \text{if } n \text{ is even,} \\ 0, & \text{if } n \text{ is odd.} \end{cases}$$

Since the sequence $1, 0, 1, 0, 1, \ldots$ does not converge, this series has no sum. We then say that the series *diverges*. We will also say that the series diverges if its partial sums tend to infinity or minus infinity.

We will need some criteria to determine if a series converges or diverges. We will also see that even knowing that a series converges, to actually determine its sum may be a very difficult problem. In any case, it will often happen that we need only determine if a series converges or not.

The first thing that can be said about a series $\sum_{k=1}^{\infty} a_k$ is that if it converges, then its terms (the a_k) tend to zero:

$$a_n = \sum_{k=1}^{n} a_k - \sum_{k=1}^{n-1} a_k = s_n - s_{n-1} \longrightarrow s - s = 0.$$

Note also that its "tails" $\sum_{k>n} a_k = s - s_n$ tend to zero... Any series whose terms do not tend to zero will be a divergent series. The opposite is not true: that the terms a_k tend to zero does not assure the convergence of a series. The typical example is the harmonic series:

$$\sum_{k=1}^{\infty} \frac{1}{k}.$$

It is clear that $\frac{1}{k} \longrightarrow 0$; however, the series diverges. Let's see a couple of proofs of this (and more will follow in Chaps. 3 and 5).

The Harmonic Series

Nicole Oresme (1323–1382) was an original and influential thinker in various disciplines, mathematics and economics among them. He was a precursor to analytic geometry, later formalized by Fermat and Descartes. He was also Bishop of Lisieux, and counselor to King Charles V of France.

Nicole gave the following proof of the divergence of the harmonic sequence. Grouping terms, write,

$$\begin{aligned}\sum_{k=1}^{\infty} \frac{1}{k} &= 1 + \frac{1}{2} + \left(\frac{1}{3} + \frac{1}{4}\right) + \left(\frac{1}{5} + \frac{1}{6} + \frac{1}{7} + \frac{1}{8}\right) + \left(\frac{1}{9} + \cdots + \frac{1}{16}\right) + \cdots \\ &> 1 + \frac{1}{2} + \frac{2}{4} + \frac{4}{8} + \frac{8}{16} + \cdots \\ &= 1 + \frac{1}{2} + \frac{1}{2} + \frac{1}{2} + \frac{1}{2} + \cdots\end{aligned}$$

but this series clearly tends to infinity. We will later see just how fast the harmonic series tends to infinity.

Another. This is by Pietro Mengoli (XVIIth Century). Bear in mind that in the harmonic series each term is the harmonic mean $(\frac{2xy}{x+y})$ of the preceding term and the following term (just as in the geometric series each term is the geometric mean $(\sqrt{xy})$ of the preceding and the following term). But the harmonic mean is always less than the arithmetic mean $(\frac{x+y}{2})$.

The harmonic mean of the numbers $\frac{1}{n-1}, \frac{1}{n}, \frac{1}{n+1}$—which is $\frac{1}{n}$—is less than its arithmetic mean $\frac{1}{3}\left(\frac{1}{n-1} + \frac{1}{n} + \frac{1}{n+1}\right)$. Therefore

$$\frac{3}{n} < \left(\frac{1}{n-1} + \frac{1}{n} + \frac{1}{n+1}\right).$$

Thus, if we suppose that $\sum_{k=1}^{\infty} \frac{1}{k}$ converges, and call S its sum,

$$\begin{aligned}
S &= 1 + \left(\frac{1}{2} + \frac{1}{3} + \frac{1}{4}\right) + \left(\frac{1}{5} + \frac{1}{6} + \frac{1}{7}\right) + \left(\frac{1}{8} + \frac{1}{9} + \frac{1}{10}\right) + \cdots \\
&> 1 + \frac{3}{3} + \frac{3}{6} + \frac{3}{9} + \cdots \\
&= 1 + \left(1 + \frac{1}{2} + \frac{1}{3} + \frac{1}{4} + \cdots\right) \\
&= 1 + S. \quad \text{Absurd, so the series diverges.}
\end{aligned}$$

Series of Positive Terms

We will restrict our attention now to *series of positive terms*, that is series $\sum_{k=1}^{\infty} a_k$ where $a_k \geq 0$ for all k. The good thing about series of positive terms is that the sequence of their partial sums is increasing: since each a_k is positive, $s_n \leq s_n + a_{n+1} = s_{n+1}$. And since the sequence s_n is increasing, by the increasing bounded sequences theorem it will be enough to see that (s_n) is a bounded sequence to verify that the series $\sum_{k=1}^{\infty} a_k$ converges. This will produce several simple criteria which ensure the convergence of a series of positive terms. First, an example.

Example The series $\sum_{k=1}^{\infty} \frac{1}{k^2}$. We may bound its partial sums as follows:

$$\begin{aligned}
s_n < s_{2^n-1} &= \sum_{k=1}^{2^n-1} \frac{1}{k^2} \\
&= \frac{1}{1^2} + \left(\frac{1}{2^2} + \frac{1}{3^2}\right) + \left(\frac{1}{4^2} + \frac{1}{5^2} + \frac{1}{6^2} + \frac{1}{7^2}\right) + \cdots + \left(\frac{1}{(2^{n-1})^2} + \cdots + \frac{1}{(2^n-1)^2}\right) \\
&\leq 1 + \left(\frac{1}{2^2} + \frac{1}{2^2}\right) + \left(\frac{1}{4^2} + \frac{1}{4^2} + \frac{1}{4^2} + \frac{1}{4^2}\right) + \cdots + \left(\frac{1}{(2^{n-1})^2} + \cdots + \frac{1}{(2^{n-1})^2}\right) \\
&= 1 + \frac{2}{2^2} + \frac{4}{4^2} + \frac{8}{8^2} + \frac{16}{16^2} + \cdots + \frac{2^{n-1}}{(2^{n-1})^2} \\
&= 1 + \frac{1}{2} + \frac{1}{4} + \frac{1}{8} + \frac{1}{16} + \cdots + \frac{1}{2^{n-1}}
\end{aligned}$$

which, as we have seen before, is less than two. Then $\sum_{k=1}^{\infty} \frac{1}{k^2}$ converges. Pietro Mengoli, whom we have already met above, asked in 1647 for the sum of this series. This became the very famous "Basel problem," solved almost ninety years later (in

1735) by Euler. It is one thing to know that a series is summable, and a very different thing to actually add it. We will see in Chap. 8 the solution to the Basel problem: $\sum_{k=1}^{\infty} \frac{1}{k^2} = \frac{\pi^2}{6}$.

A general comment before going into the criteria: the convergence or divergence of a series will never be affected by what happens with finitely many of its terms. Finitely many terms can always be added. Thus, if the following criteria are satisfied only by a tail ($\sum_{k>N} a_k$) of the series, this will be enough to determine its convergence. The first N terms of a series never affect its convergence.

Comparison Criterion If for all k, $a_k \leq cb_k$ and the series $\sum_{k=1}^{\infty} b_k$ converges, then the series $\sum_{k=1}^{\infty} a_k$ converges.

$$s_n = \sum_{k=1}^{n} a_k \leq c \sum_{k=1}^{n} b_k \leq c \sum_{k=1}^{\infty} b_k,$$

so the sequence s_n is bounded (and increasing). Thus $\sum_{k=1}^{\infty} a_k$ converges.

Example The series $\sum_{k=1}^{\infty} \frac{1}{2^k+k}$. For each k we have $\frac{1}{2^k+k} < \frac{1}{2^k}$, and we know that $\sum_{k=1}^{\infty} \frac{1}{2^k}$ converges, so $\sum_{k=1}^{\infty} \frac{1}{2^k+k}$ also converges.

Root Criterion *If for all k, $\sqrt[k]{a_k} \leq r < 1$, then the series $\sum_{k=1}^{\infty} a_k$ converges.*

We have for each k,

$$a_k \leq r^k,$$

so taking $b_k = r^k$ and using the comparison criterion, the series $\sum_{k=1}^{\infty} a_k$ is convergent.

Example The series $\sum_{k=1}^{\infty} \frac{k^k}{(2k+1)^k}$. We have

$$\sqrt[k]{a_k} = \sqrt[k]{\frac{k^k}{(2k+1)^k}} = \frac{k}{(2k+1)} \leq \frac{1}{2} < 1,$$

thus the series $\sum_{k=1}^{\infty} \frac{k^k}{(2k+1)^k}$ converges.

Quotient Criterion *If for all k, $\frac{a_{k+1}}{a_k} \leq r < 1$, then the series $\sum_{k=1}^{\infty} a_k$ converges.*

We have for each k,

$$a_{k+1} \leq ra_k \leq r^2 a_{k-1} \leq r^3 a_{k-2} \leq \cdots \leq r^k a_1,$$

so taking $b_{k+1} = r^k$, since $\sum_{k=1}^{\infty} r^k$ converges, we can use the comparison criterion to obtain that $\sum_{k=1}^{\infty} a_k$ converges.

Example The series $\sum_{k=1}^{\infty} \frac{1}{k!}$. We have, for each $k \geq 1$,

$$\frac{a_{k+1}}{a_k} = \frac{k!}{(k+1)!} = \frac{1 \cdot 2 \cdot 3 \cdots k}{1 \cdot 2 \cdot 3 \cdots k \cdot (k+1)} = \frac{1}{k+1} \leq \frac{1}{2} < 1,$$

thus the series $\sum_{k=1}^{\infty} \frac{1}{k!}$ converges. In Chap. 3 we will see its sum.

Series with Positive and Negative Terms

If a series has infinitely many positive terms, but also infinitely many negative terms, it will be convenient for our discussion to distinguish between them. To this end, we define

$$a_k^+ = \begin{cases} a_k, & \text{if } a_k > 0 \\ 0, & \text{if } a_k \leq 0, \end{cases}$$

and

$$a_k^- = \begin{cases} 0, & \text{if } a_k \geq 0 \\ a_k, & \text{if } a_k < 0. \end{cases}$$

For example, for the series $\sum_{k=1}^{\infty} \frac{(-1)^k}{k}$, this is:

$$a_k : -1, \frac{1}{2}, \frac{-1}{3}, \frac{1}{4}, \frac{-1}{5}, \cdots$$

$$a_k^+ : 0, \frac{1}{2}, 0, \frac{1}{4}, 0, \cdots$$

$$a_k^- : -1, 0, \frac{-1}{3}, 0, \frac{-1}{5}, \cdots$$

$$|a_k| : 1, \frac{1}{2}, \frac{1}{3}, \frac{1}{4}, \frac{1}{5}, \cdots$$

Note that $a_k^+ - a_k^- = |a_k|$, the absolute value of a_k (the same we used when we talked about distance). Note also that

$$\sum_{k=1}^{n} a_k = \sum_{k=1}^{n} a_k^+ + \sum_{k=1}^{n} a_k^-, \qquad \text{and}$$

$$\sum_{k=1}^{n} |a_k| = \sum_{k=1}^{n} a_k^+ - \sum_{k=1}^{n} a_k^- .$$

We will say that a series $\sum_{k=1}^{\infty} a_k$ is *absolutely convergent* if the series (of positive terms) $\sum_{k=1}^{\infty} |a_k|$ is convergent. "Absolute" convergence is stronger than convergence, in the following sense: if the series $\sum_{k=1}^{\infty} |a_k|$ converges, then the series $\sum_{k=1}^{\infty} a_k$ also converges. The reason for this is the following: using the comparison criterion,

$$\text{since } a_k^+ \leq |a_k| \text{ for all } k, \sum_{k=1}^{\infty} a_k^+ \text{ converges, and}$$

$$\text{since } -a_k^- \leq |a_k| \text{ for all } k, \sum_{k=1}^{\infty} -a_k^- \text{ converges, and is } -\sum_{k=1}^{\infty} a_k^-,$$

so $\sum_{k=1}^{\infty} a_k^-$ converges. But then

$$\sum_{k=1}^{n} a_k = \sum_{k=1}^{n} a_k^+ + \sum_{k=1}^{n} a_k^- \longrightarrow \sum_{k=1}^{\infty} a_k^+ + \sum_{k=1}^{\infty} a_k^- .$$

Note that

$$\sum_{k=1}^{n} |a_k| = \sum_{k=1}^{n} a_k^+ - \sum_{k=1}^{n} a_k^-,$$

so $\sum_{k=1}^{n} |a_k|$ converges if and only if both $\sum_{k=1}^{n} a_k^+$ and $\sum_{k=1}^{n} a_k^-$ converge.

But a series $\sum_{k=1}^{\infty} a_k$ may converge without doing so absolutely, that is, without $\sum_{k=1}^{\infty} |a_k|$ converging. The following convergence criterion for "alternating" series will provide us with an example.

Leibniz Criterion *If a_k are positive numbers which decrease to zero ($0 < \cdots \leq a_{k+1} \leq a_k \leq \cdots$ and $\lim a_k = 0$), then the series*

$$\sum_{k=1}^{\infty} (-1)^k a_k = -a_1 + a_2 - a_3 + a_4 - a_5 + \cdots$$

converges.

Let's see why: call the partial sums $s_n = \sum_{k=1}^{n} (-1)^k a_k$. Since

$$s_n = s_{n-1} + (-1)^n a_n,$$

each term added to the partial sums produces a jump to the right—if n is even—or to the left—if n is odd—. Also, since the a_k decrease, these jumps get smaller and smaller... We then have

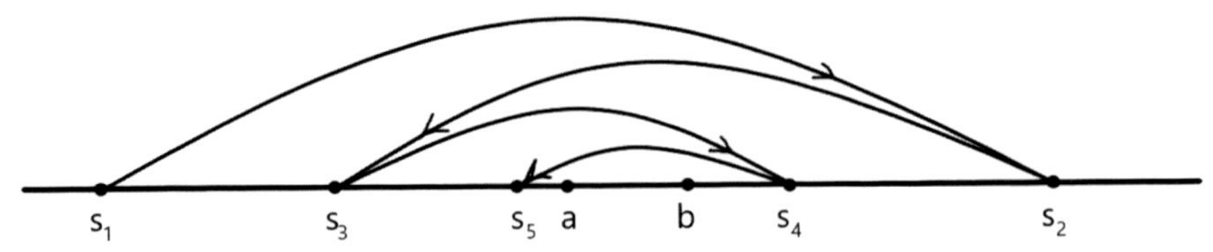

So, the odd partial sums form an increasing sequence (and are bounded by s_2), while the even partial sums form a decreasing sequence (and are bounded by s_1). By the increasing bounded sequences theorem, they both converge. Say the odd partial sums tend to a, and the even ones to b ($a \leq b$). But for all n, $b - a \leq s_{2n} - s_{2n-1} = a_{2n}$, which tends to zero. Thus $a = b$ and the partial sums (the odd and the even) converge to the same number. The series is convergent. □

Note also that $|s_n - a| < a_n$, in other words, the distance between a partial sum and the limit is smaller than the last term added.

Example Gregory's series $\sum_{k=0}^{\infty} \frac{(-1)^k}{2k+1} = 1 - \frac{1}{3} + \frac{1}{5} - \frac{1}{7} + \cdots$. It converges by Leibniz' criterion, for $a_k = \frac{1}{2k+1}$ decrease to zero. In Chap. 5 we will see that its sum is $\frac{\pi}{4}$.

Example The alternating harmonic series $\sum_{k=1}^{\infty}(-1)^k \frac{1}{k}$. It converges, for $a_k = \frac{1}{k}$ decrease to zero. We will see in Chap. 6 that it adds $-\ln 2$.

This is also an example of a convergent series that is *not* absolutely convergent because—as we have seen—$\sum_{k=1}^{\infty} \frac{1}{k}$ diverges.

The Riemann Series Theorem

Bernhard Riemann (1826–1866) was a German mathematician. He gave the first formal definition of the integral and also made important contributions to complex analysis and differential geometry. His work on analytic number theory includes the formulation of what is today known as the Riemann hypothesis, perhaps the most important still unsolved problem in Mathematics.

When we add finitely many numbers we know that altering the order in which we add these numbers will not change their sum. But when we have infinitely many terms we will have to be more careful, as the following surprising result of Riemann shows. In its proof we will use an idea similar to that of the proof of the Leibniz criterion:

The Riemann Series Theorem *If the series $\sum_{k=1}^{\infty} a_k$ converges, but* not *absolutely, by changing the order of its terms a_k we can obtain as its sum,* any *real number we want.*

Here's the proof: First, note that since $\sum_{k=1}^{\infty} a_k$ converges, $a_k \longrightarrow 0$. On the other hand, since $\sum_{k=1}^{\infty} |a_k|$ diverges, both the series of its positive terms $\sum_{k=1}^{\infty} a_k^+$ and that of its negative terms $\sum_{k=1}^{\infty} a_k^-$ diverge: If both were convergent, so would be $\sum_{k=1}^{\infty} |a_k|$, if one were convergent but not the other, since $\sum_{k=1}^{n} a_k = \sum_{k=1}^{n} a_k^+ + \sum_{k=1}^{n} a_k^-$, $\sum_{k=1}^{\infty} a_k$ would not converge. Thus, $\sum_{k=1}^{n} a_k^+$ tends to ∞ with n, while $\sum_{k=1}^{n} a_k^-$ tends to $-\infty$.

Take any real number c, and let's say we want the sum of the terms of our series to be c. To that end, we will reorder the series as follows: we will add alternately blocks of positive terms and blocks of negative terms

$$a_1^+, a_2^+, a_3^+, a_4^+, a_5^+, \cdots$$
$$a_1^-, a_2^-, a_3^-, a_4^-, a_5^-, \cdots$$

to approximate c. First, add just enough positive terms to overcome c—say, p_1 terms—that is:

$$\sum_{k=1}^{p_1-1} a_k^+ \leq c < \sum_{k=1}^{p_1} a_k^+.$$

Now, add just enough negative terms (say q_1) to pass to the left of c:

$$\sum_{k=1}^{p_1} a_k^+ + \sum_{k=1}^{q_1} a_k^- < c \leq \sum_{k=1}^{p_1} a_k^+ + \sum_{k=1}^{q_1-1} a_k^-.$$

Now just enough (say, p_2) to overcome c again:

$$\sum_{k=1}^{p_1} a_k^+ + \sum_{k=1}^{q_1} a_k^- + \sum_{k=p_1+1}^{p_2-1} a_k^+ \leq c < \sum_{k=1}^{p_1} a_k^+ + \sum_{k=1}^{q_1} a_k^- + \sum_{k=p_1+1}^{p_2} a_k^+.$$

And then just as many negative terms as is necessary to be on the left of c again, etc. What will this reordered series add? At each step, the difference between the partial sum and c is smaller than the last number added. But as $a_k \longrightarrow 0$, this difference becomes smaller and smaller... as small as we want. □

For example, by conveniently reordering

$$-1, \frac{1}{2}, \frac{-1}{3}, \frac{1}{4}, \frac{-1}{5}, \frac{1}{6}, \frac{-1}{7}, \cdots$$

we can have them sum $\sqrt{2}$, or π... or whatever.

Absolute and Unconditional Convergence

A series $\sum_{k=1}^{\infty} a_k$ is said to be *unconditionally convergent* if it converges to the same sum regardless of the order in which its terms are added. By the Riemann series Theorem we have just seen, such a series will be absolutely convergent. We see now that the converse is also true.

Theorem *If a series converges absolutely, its terms commute.*

The proof is the following: Let's say that the series $\sum_{k=1}^{\infty} a_k$ converges absolutely, and that its sum is s, that is

$$s_n = \sum_{k=1}^{n} a_k \longrightarrow s \text{ para } n \to \infty.$$

Consider now any reordering $a'_1, a'_2, \ldots$ of the terms of the series. We want to check that

$$s'_n = \sum_{k=1}^{n} a'_k \longrightarrow s \text{ (the same as before!) for } n \to \infty.$$

To see this, take $\varepsilon > 0$. By the absolute convergence of the series, we will have, for a sufficiently large N,

$$\sum_{k>N} |a_k| < \varepsilon.$$

Consider then M so large that the set $\{a'_1, a'_2, \ldots, a'_M\}$ contains the set $\{a_1, a_2, \ldots, a_N\}$. Now for all $n > M$

$$|s - s'_n| = \left| \sum_{k=1}^{\infty} a_k - \sum_{k=1}^{n} a'_k \right| \leq \sum_{k>N} |a_k| < \varepsilon,$$

for in omitting $\{a'_1, a'_2, \ldots, a'_M, \ldots, a'_n\}$ we are undoubtedly also omitting $\{a_1, a_2, \ldots, a_N\}$. Hence, $s'_n \longrightarrow s$. □

Exercises

1 Prove that $s = \sup A$ if and only if s is an upper bound of A and there is a sequence (a_n) of elements of A such that $a_n \longrightarrow s$.

2 Prove that if $a_n \longrightarrow c$, $b_n \longrightarrow c$, and $a_n \leq c_n \leq b_n$ for all n, then $c_n \longrightarrow c$.

3 Prove that if $a_n \longrightarrow a$ and $b_n \longrightarrow b$,

(i) $a_n + b_n \longrightarrow a + b$.
(ii) $a_n b_n \longrightarrow ab$.

4 Given a sequence (x_n), bounded from above and below, consider its *tails*:

$$A_n = \{x_k : k \geq n\}.$$

(i) Prove that for all n we have $A_n \supset A_{n+1}$.
(ii) Determine the relationship between

$$s_n = \sup A_n \qquad i_n = \inf A_n \qquad s_{n+1} = \sup A_{n+1} \qquad i_{n+1} = \inf A_{n+1}.$$

iii) Prove that (i_n) converges and that (s_n) converges.
iv) If $i_n \longrightarrow i$ and $s_n \longrightarrow s$, what is the order relation between i and s?

The number $i = \liminf x_n$ is called the *limit inferior* of x_n, and $s = \limsup x_n$ is called the *limit superior* of x_n.

5 Prove that the following are equivalent:

(i) $L = \limsup x_n$.
(ii) for all $\varepsilon > 0$:
   (a) there are infinitely many n such that $L - \varepsilon < x_n$, and
   (b) there are at most finitely many n such that $L + \varepsilon < x_n$.

6 We will say that (x_n) is a *Cauchy* sequence if for any $\varepsilon > 0$ there exists an n_ε such that if $m, n \geq n_\varepsilon$, then

$$|x_m - x_n| < \varepsilon.$$

Prove that all Cauchy sequences are bounded.

7 Prove that the following are equivalent:

(i) (x_n) converges.
(ii) $\liminf x_n = \limsup x_n$.
(iii) (x_n) is Cauchy.

8 Which of the following implications hold? (Give a proof or a counterexample)

(a) $a_{n+1} - a_n \longrightarrow 0 \Rightarrow (a_n)$ converges.
(b) $|a_m - a_n| < \frac{1}{n+m}$ for all $n, m \Rightarrow (a_n)$ converges.
(c) $|a_{n+1} - a_n| < \frac{1}{2^n}$ for all $n \Rightarrow (a_n)$ converges.
(d) $a_{n+1} - a_n \longrightarrow 0 \Rightarrow a_{2n} - a_n \longrightarrow 0$.

9 An exercise about *tails* of a series.

(a) Given the geometric series $\sum_{k=0}^{\infty} r^k$, (with $|r| < 1$), prove that for each $n \in \mathbb{N}$ nth tail is

$$\sum_{k>n} r^k = \frac{r^{n+1}}{r}. \qquad \text{For example, for } r = \frac{1}{2}, \text{ is } \sum_{k>n} \frac{1}{2^k} = \frac{1}{2^n}.$$

(b) Given the alternating series $\sum_{k=1}^{\infty}(-1)^k a_k$, (with a_k positive and decreasing),

$$\left|\sum_{k>n}(-1)^k a_k\right| = |s - s_n| \leq |a_n|.$$

10 Which of the following implications hold? (Give a proof or a counterexample)

(a) $\sum_n a_n < \infty$ and $nb_n < a_n$ for all $n \Rightarrow \sum_n b_n < \infty$.
(b) $\sum_n a_n = \infty$ and $nb_n > a_n$ for all $n \Rightarrow \sum_n b_n = \infty$.

11 If $\frac{a_n}{b_n} \longrightarrow L$, what can be said about the convergence of $\sum_n a_n$,

(i) if $\sum_n b_n$ converges?
(ii) if $\sum_n b_n$ diverges?

(consider the cases $L = 0$, $L > 0$, and $L = \infty$).

12 It is known that the series $\sum_n a_n$ converges, but not absolutely. Reorder it so that $\sum_n a_n = \infty$.

3 Functions

We consider here the elementary functions, and also curves in the plane and their parametrization. We introduce the notion of continuity, and present the theorems of Bolzano and of Weierstrass.

The Elementary Functions

What is the surface area of a square whose sides measure two? Four. And if those sides measure three? Nine. And if they measure x? x^2. This last question (and its answer) is of a very different type from the first two. It is more abstract. For it refers to the side of a square whose measure is x—a generic number, in other words, a number of which we know nothing. But this does not hinder us from giving an—equally generic—answer. The answers to the first two questions give us the area of squares of sides two or three. The answer to the more abstract question allows us to know the area of any square. This is infinitely more useful. Abstraction is what confers usefulness to mathematics.

When to any "generic" number x we assign (in whatever way) another number, what we have is a *function*. If we call f the function of the previous paragraph, then we will write

$$f(x) = x^2 \quad \text{or} \quad x \mapsto x^2$$

to express that each number x (usually called the *variable*) is assigned the number x^2 by our function f.

We will return later to the abstract idea of function. But first we should consider some of the functions known as "elementary functions," the simplest: polynomials, the circular functions (sine and cosine), and the exponential function.

I. Zalduendo, *Calculus off the Beaten Path*, SUMS Readings,
https://doi.org/10.1007/978-3-031-15765-3_3

Polynomials

After the "constant" functions (which assign to all numbers x the same number c), perhaps the simplest are those that are obtained from the variable x by addition and multiplication; for example,

$$x \mapsto x^2, \quad x \mapsto x^2 + 3x, \quad x \mapsto 4x^3 - \frac{1}{2}x + 5,$$

these are the *polynomials*, in general, the functions of the form

$$f(x) = a_n x^n + a_{n-1} x^{n-1} + \cdots + a_2 x^2 + a_1 x + a_0,$$

where each a_i is a fixed real number, independent of x. In this expression, n is the *degree* of the polynomial. Note that for each value of x, the value of the function, $f(x)$, is obtained by summing and multiplying. Many of the other commonly used functions are not so easy to calculate, and for this reason one of the objectives of Calculus is to approximate these other functions by means of polynomials.

On the other hand, the simplicity of polynomials makes them "rigid" in the following sense: consider the polynomials of degree one

$$f(x) = ax + b,$$

and say that we wish to find one such that $f(1) = -2$ and $f(3) = 4$, that is:

$$\begin{aligned} f(1) =& a + b = -2 \\ f(3) =& 3a + b = 4. \end{aligned}$$

We may solve this system of two linear equations with two unknowns and obtain (with $a = 3$ and $b = -5$) the polynomial we want:

$$f(x) = 3x - 5.$$

But we cannot ask more of this polynomial (for example, that $f(2) = 0 \ldots$). In general, to be able to ask more, to impose more conditions, we must permit the degree of the polynomial to increase. The greater the degree, the lesser the rigidity, and the greater the possibility of adjusting data or approximating a given function. We will come back on this matter when we consider Taylor polynomials in Chap. 6.

Circular Functions

It is one thing to define the sine or the cosine of an angle α, and another to define the sine function or the cosine function. Because the "angle α" is a geometric object,

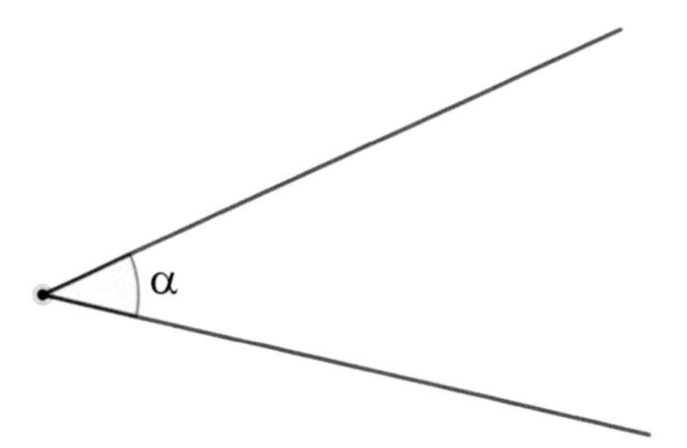

but a function must have as its variable a number, not a geometric object. To identify the angle α with a number, one thing that one can do is measure it. And there are several ways of doing this, because our angle α can be measured with different units: *sexagesimal degrees* (in which the right angle measures $90°$), *centesimal degrees* (in which the right angle measures $100°$), or *radians* (in which the right angle measures $\frac{\pi}{2}$). And these different ways of measuring the angle α give rise to different "cosine" functions and different "sine" functions, as you may have noticed if you've used your calculator a little carelessly.

The circular functions—cosine and sine—which we will define correspond to measuring angles with radians. Consider in the plane the circle with center at the origin (the point $(0, 0)$) and of radius one. Given the number t, we measure from the point $(1, 0)$ an arc of length t on the circumference (upwards if t is positive, downwards if t is negative). We thus reach a point $p(t)$ on the circumference:

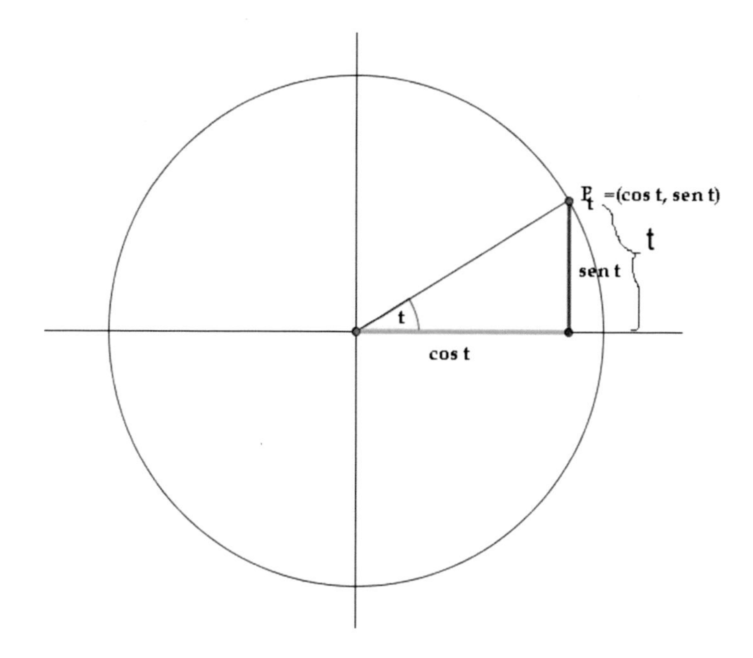

The point $p(t)$ has two coordinates. The first we will call *cosine of t* and the second *sine of t*: that is $p(t) = (\cos t, \sin t)$. To each real number t corresponds a point $p(t)$ (if t is larger than 2π which is the measure of the complete circumference, we continue going around the circle). Thus we have defined the "circular functions" $\cos t$ and $\sin t$ for any value of t. Several things about these functions can be seen by simply looking at the above graph. For example,

$$\cos(-t) = \cos t$$
$$\sin(-t) = -\sin t$$
$$\sin(\pi - t) = \sin t$$
$$\cos(\pi - t) = -\cos t$$
$$\cos\left(\frac{\pi}{2} - t\right) = \sin t$$
$$\sin\left(\frac{\pi}{2} - t\right) = \cos t$$
$$\text{and, } \cos^2 t + \sin^2 t = 1.$$

Note that the usual "triangular" definitions of sine and cosine as relations between the sides of a right triangle may be derived from the same picture. The hypotenuse here measures one (the radius), and relations between the sides are the same for all similar (same *shape*) triangles.

Later on, we will need the following formulas for sine and cosine of a sum:

$$\sin(x + y) = \cos x \sin y + \cos y \sin x$$
$$\text{and} \quad \cos(x + y) = \cos x \cos y - \sin y \sin x.$$

Here is a simple proof for these formulas. Let's start with the sine of a sum. For this, calculate the area of the following triangle:

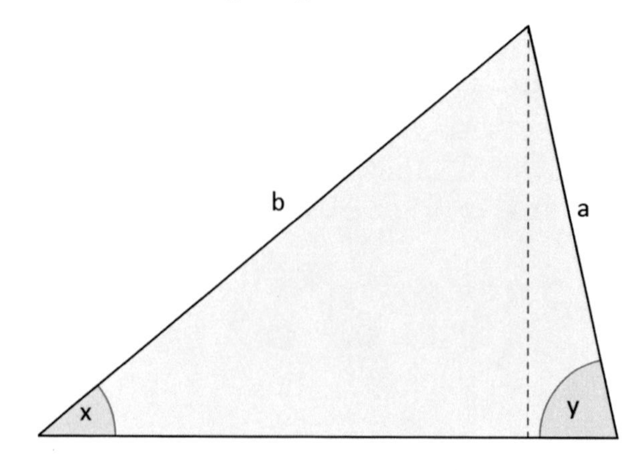

$$\text{Area } = \frac{1}{2} \cdot b \cos x \cdot a \sin y + \frac{1}{2} \cdot a \cos y \cdot b \sin x = \frac{ab}{2} [\cos x \sin y + \cos y \sin x] .$$

Now rotate and calculate the area again:

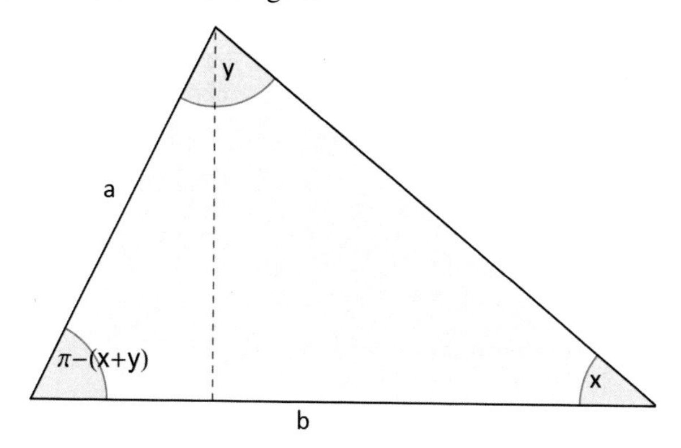

$$\text{Area } = \frac{1}{2} \cdot b \cdot a \sin(\pi - (x + y)) = \frac{ab}{2} [\sin(x + y)] .$$

Then $\sin(x + y) = \cos x \sin y + \cos y \sin x$. For the cosine of a sum,

$$\begin{aligned} \cos(x + y) &= \sin\left(\frac{\pi}{2} - (x + y)\right) \\ &= \sin\left(\left(\frac{\pi}{2} - x\right) + (-y)\right) \\ &= \sin\left(\frac{\pi}{2} - x\right) \cos(-y) + \sin(-y) \cos\left(\frac{\pi}{2} - x\right) \\ &= \cos x \cos y - \sin y \sin x . \end{aligned}$$

Done!

The Exponential Function: Bernoulli's Inequality

Leonhard Euler (1707–1783) was born in Basel. His father was a friend of Johann Bernoulli, one of the foremost mathematicians of the time, who took an interest in Leonhard's mathematical education. By age twenty he was working in the recently created Saint Petersburg Academy of Science. Some years later (1740) he moved to the Berlin Academy where he worked most of his life (before returning to St. Petersburg). Although he became blind in 1765, he continued working up to his death in 1783. Euler was a most prolific mathematician. He published works on number theory, differential equations, calculus of variations, and graph theory. The number e, which we will define below, is designated with that letter in his honor.

The exponential function is surely the most important function in mathematics. Beyond its theoretical importance, it is present in countless practical applications. Let's begin with one of them, compound interest: you put money in the bank, and the bank promises to pay you 6% annual interest. By the end of the year, for each dollar invested the bank will give you 6 cents, thus your capital will be multiplied by $(1 + 0,06)$. But it would be better for you if the bank gave you 3% every six months; then the first 3 cents would become part of your capital and would earn interest during the second semester. By the end of the year your capital would be multiplied by

$$(1 + 0,03)(1 + 0,03) = \left(1 + \frac{0,06}{2}\right)^2 .$$

And if capitalized every four months (three periods),

$$(1 + 0,02)(1 + 0,02)(1 + 0,02) = \left(1 + \frac{0,06}{3}\right)^3 ,$$

and if every three months (four periods),

$$\left(1 + \frac{0,06}{4}\right)^4 .$$

If capitalization occurred instantly (infinitely many infinitely short periods), by year's end your capital would be multiplied by the number

$$\lim_{n \longrightarrow \infty} \left(1 + \frac{0,06}{n}\right)^n .$$

Now, you may object, no bank will give me instantaneous interest. True. But Nature apparently will: many natural processes involve growth which is proportional, at any given moment, to the amount present. Thus, it will be important to calculate the limit

$$\lim_{n \longrightarrow \infty} \left(1 + \frac{x}{n}\right)^n ,$$

which is what we will do now. Say—for the moment—that the growth rate is a fixed number x, and for every natural number n, set

$$c_n = \left(1 + \frac{x}{n}\right)^n .$$

We want to see if the sequence c_n converges, and to what. We will use "Newton's formula":

$$(a+b)^n = \sum_{k=0}^{n} \binom{n}{k} a^{n-k} b^k, \quad \text{where } \binom{n}{k} = \frac{n!}{(n-k)!k!}$$

Let's begin:

$$\begin{aligned}
c_n &= \left(1+\frac{x}{n}\right)^n \\
&= \sum_{k=0}^{n} \binom{n}{k} \left(\frac{x}{n}\right)^k && \text{(using Newton's formula)} \\
&= \sum_{k=0}^{n} \frac{x^k}{k!} \frac{n!}{(n-k)!n^k} \\
&= \sum_{k=0}^{n} \frac{x^k}{k!} \left[\frac{n\cdot(n-1)\cdots(n-k+1)}{n^k}\right] && \text{(simplifying the factorials)} \\
&= \sum_{k=0}^{n} \frac{x^k}{k!} \left[\frac{n}{n}\cdot\frac{n-1}{n}\cdots\frac{n-(k-1)}{n}\right] \\
&= \sum_{k=0}^{n} \frac{x^k}{k!} \left[\left(1-\frac{1}{n}\right)\cdots\left(1-\frac{k-1}{n}\right)\right] \\
&< \sum_{k=0}^{n} \frac{x^k}{k!} \\
&< \sum_{k=0}^{\infty} \frac{x^k}{k!},
\end{aligned}$$

but this series converges absolutely by the quotient criterion:

$$\frac{|x|^{k+1}}{(k+1)!} \cdot \frac{k!}{|x|^k} = \frac{|x|}{k+1} < 1 \quad \text{as soon as } k > |x|-1.$$

thus the sequence c_n is bounded. But also, two things have appeared which deserve our attention:

(a) the product $\left[\left(1-\frac{1}{n}\right)\cdots\left(1-\frac{k-1}{n}\right)\right]$, for which we will use the notation

$$\prod_{j=1}^{k-1}\left(1-\frac{j}{n}\right),$$

analogous to the summation notation $\sum$, but for multiplication,

(b) the function

$$f(x) = \sum_{k=0}^{\infty} \frac{x^k}{k!}.$$

If we consider very large n, the product in a) will be close to one, and in our discussion above the c_n's seem to converge to the function $f(x)$ in (b). This will indeed be so. But we will need to consider the product closely, and for this we will use Bernoulli's inequality.

Jacob Bernoulli (1655–1705) is, with his smaller brother Johann, the two most notable in a family of extraordinary mathematicians. Jacob made important contributions to the theory of differential equations, and to the theory of probability.

Bernoulli's Inequality *Given $a_1, \dots, a_n$, either all positive or all between -1 and 0, then*

$$\prod_{i=1}^{n} (1 + a_i) \geq 1 + \sum_{i=1}^{n} a_i.$$

We will prove this inequality by induction. For $n = 1$ we have

$$1 + a_1 \geq 1 + a_1.$$

Now, supposing the inequality valid for n,

$$\begin{aligned} \prod_{i=1}^{n+1} (1 + a_i) &= \prod_{i=1}^{n} (1 + a_i)\,(1 + a_{n+1}) \\ &\geq \left(1 + \sum_{i=1}^{n} a_i\right)(1 + a_{n+1}) \\ &= 1 + \sum_{i=1}^{n} a_i + a_{n+1} + a_{n+1} \sum_{i=1}^{n} a_i \\ &\geq 1 + \sum_{i=1}^{n+1} a_i, \end{aligned}$$

where in the first inequality we have used that $1 + a_{n+1} \geq 0$, and in the second that all a_i's have the same sign, and thus $a_{n+1} \sum_{i=1}^{n} a_i \geq 0$. □

Note that if the a_i's are all the same, we have: for $-1 < a$,

$$(1+a)^n \geq 1 + na,$$

which is also usually called Bernoulli's inequality.

Now set $s_n = \sum_{k=0}^{n} \frac{x^k}{k!}$, the partial sum of the series in b), and calculate its difference with c_n. Bearing in mind that the first two terms coincide in s_n and c_n, we sum for $k \geq 2$:

$$s_n - c_n = \sum_{k=2}^{n} \frac{x^k}{k!}\left(1 - \left[\left(1-\frac{1}{n}\right)\cdots\left(1-\frac{k-1}{n}\right)\right]\right)$$

using Bernoulli's inequality,

$$\begin{aligned}
&\leq \sum_{k=2}^{n} \frac{x^k}{k!}\left(1 - \left[1 - \sum_{j=1}^{k-1}\frac{j}{n}\right]\right) \\
&= \sum_{k=2}^{n} \frac{x^k}{k!}\left(\frac{1}{n}\sum_{j=1}^{k-1} j\right) \\
&= \sum_{k=2}^{n} \frac{x^k}{k!}\frac{1}{n}\frac{k(k-1)}{2} && \text{(sum of the first } k-1 \text{ natural numbers)} \\
&\leq \sum_{k=2}^{\infty} \frac{x^k}{(k-2)!}\frac{1}{2n} \\
&= \sum_{k=0}^{\infty} \frac{x^k}{k!}\frac{x^2}{2n} && \text{(renaming indices)} \\
&= \frac{x^2}{2n}\sum_{k=0}^{\infty} \frac{x^k}{k!},
\end{aligned}$$

which tends to zero, for the series converges and $\frac{x^2}{2n} \longrightarrow 0$ as n grows.

Thus $\lim_n c_n = \lim_n s_n$:

$$\lim_{n\longrightarrow\infty}\left(1+\frac{x}{n}\right)^n = \sum_{k=0}^{\infty}\frac{x^k}{k!} = f(x).$$

We will call this function f, defined either as the limit of $(1+\frac{x}{n})^n$ or as the series $\sum_{k=0}^{\infty}\frac{x^k}{k!}$, the *exponential function*. We will now see some of its properties, which justify the name.

(i) $f(x+y) = f(x)f(y)$, because:

$$f(x+y) = \sum_{k=0}^{\infty} \frac{(x+y)^k}{k!}$$

$$= \sum_{k=0}^{\infty} \sum_{j=0}^{k} \frac{1}{k!} \binom{k}{j} x^j y^{k-j}.$$

As the series converges absolutely, we may reorder its terms. We will do this in the following way. As is, if we consider the index pairs (k, j), we are summing along the columns, from left to right in

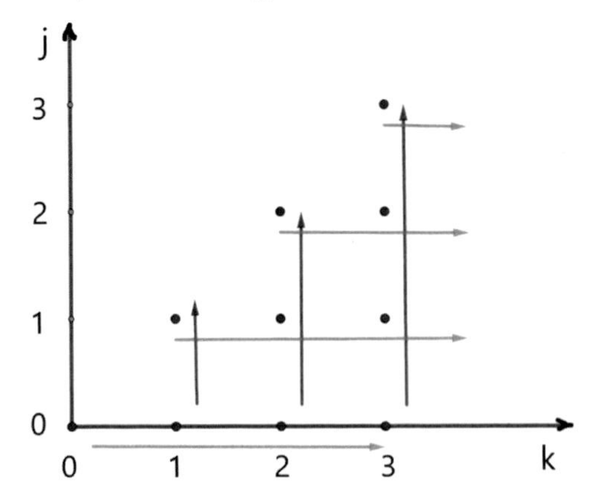

Instead, let's sum along the rows, from the bottom up:

$$= \sum_{j=0}^{\infty} \sum_{k=j}^{\infty} \frac{1}{k!} \binom{k}{j} x^j y^{k-j}$$

$$= \sum_{j=0}^{\infty} \sum_{k=j}^{\infty} \frac{1}{k!} \frac{k!}{(k-j)!j!} x^j y^{k-j}$$

$$= \sum_{j=0}^{\infty} \frac{x^j}{j!} \sum_{k=j}^{\infty} \frac{y^{k-j}}{(k-j)!}$$

$$= \sum_{j=0}^{\infty} \frac{x^j}{j!} \sum_{i=0}^{\infty} \frac{y^i}{i!} \qquad \text{(calling } k-j, i\text{)}$$

$$= f(x)f(y).$$

(ii) f is always strictly positive, for $f(x) = f(\frac{x}{2} + \frac{x}{2}) = f(\frac{x}{2})^2 \geq 0$, but if there were an a such that $f(a) = 0$, we would have $f(0) = f(a + (-a)) = f(a)f(-a) = 0$, but this does not happen, because

(iii) $f(0) = 1$.

We will call "e" the number $f(1)$. That is,

$$e = \lim_{n \longrightarrow \infty} \left(1 + \frac{1}{n}\right)^n \quad \text{and} \quad e = \sum_{k=0}^{\infty} \frac{1}{k!}.$$

Note that

$$\begin{aligned} e^x &= \lim_{n \to \infty} \left(1 + \frac{1}{n}\right)^{nx} \\ &= \lim_{n \to \infty} \left(1 + \frac{x}{nx}\right)^{nx} \\ &= \lim_{p \to \infty} \left(1 + \frac{x}{p}\right)^{p} \\ &= f(x), \end{aligned}$$

so we will use the notation e^x for the exponential function. In other words

$$e^x = \lim_{n \longrightarrow \infty} \left(1 + \frac{x}{n}\right)^n \quad \text{and} \quad e^x = \sum_{k=0}^{\infty} \frac{x^k}{k!}.$$

The number e is one of the most important in mathematics. In decimal notation, something like

$$e = 2,718\ldots$$

Irrationality of *e*

The number *e* is irrational.

To prove this, we suppose it is rational and will reach a contradiction. Write then $e = \frac{m}{n}$. Thus $\frac{n}{m} = e^{-1} = \sum_{k=0}^{\infty} \frac{(-1)^k}{k!}$, and

$$n(m-1)! = \frac{n}{m} m!$$

$$= \sum_{k=0}^{\infty} \frac{(-1)^k}{k!} m!$$

$$= \sum_{k=0}^{m} \frac{(-1)^k}{k!} m! \pm \frac{1}{(m+1)} \mp \frac{1}{(m+1)(m+2)} \pm \cdots$$

Then,

$$n(m-1)! - \sum_{k=0}^{m} \frac{(-1)^k}{k!} m! =$$

$$= \pm \frac{1}{(m+1)} \mp \frac{1}{(m+1)(m+2)} \pm \frac{1}{(m+1)(m+2)(m+3)} \mp \cdots$$

Now, on the left-hand side of the equality we have a whole number, call it p, and on the right-hand side, an alternating series whose sum we know, from our proof of the Leibniz criterion, is a number between the first term and the sum of the first two. Thus,

$$\frac{1}{(m+1)} - \frac{1}{(m+1)(m+2)} < |p| < \frac{1}{(m+1)}$$

$$\frac{1}{(m+2)} < |p| < \frac{1}{(m+1)},$$

but this cannot be, for p is a whole number. We conclude then that e must be irrational. □

Convergence of $\prod_{k=1}^{\infty}(1 + a_k)$ and of $\sum_{k=1}^{\infty} a_k$

Note that since $e^x = \sum_{k=0}^{\infty} \frac{x^k}{k!} = 1 + x + \sum_{k=2}^{\infty} \frac{x^k}{k!}$, it is clear that

$$1 + x \leq e^x \text{ for all } x \geq 0.$$

We will use this to prove the following:

Proposition *If $a_k \geq 0$ for all k (or if $-1 < a_k < 0$ for all k), $\prod_{k=1}^{\infty}(1 + a_k)$ converges if and only if $\sum_{k=1}^{\infty} a_k$ converges*

Here $\prod_{k=1}^{n}(1 + a_k)$ means the product $(1 + a_1)(1 + a_2) \ldots (1 + a_n)$ and that the infinite product "converges" means there exists a limit to the "partial products" as $n \to \infty$. Note that if $a_k \geq 0$ for all k, both the partial sums $\sum_{k=1}^{n} a_k$ and the partial products $\prod_{k=1}^{n}(1 + a_k)$ form increasing sequences (while if $-1 < a_k < 0$ for all k, they are decreasing). Thus to check their convergence it will be enough to see that they are bounded. For each n we have

$$1+\sum_{k=1}^{n} a_k \leq \prod_{k=1}^{n}(1+a_k) \leq \prod_{k=1}^{n} \mathrm{e}^{a_k} = \mathrm{e}^{\sum_{k=1}^{n} a_k}.$$

Where the first inequality is Bernoulli's, and the second is because $1+x \leq \mathrm{e}^x$. Therefore, if $\prod_{k=1}^{\infty}(1+a_k)$ converges, $s_n = \sum_{k=1}^{n} a_k$ is bounded, while if $\sum_{k=1}^{\infty} a_k$ converges, $p_n = \prod_{k=1}^{n}(1+a_k)$ is bounded. □

For example, if $a_k = \frac{1}{k}$ noting that

$$\prod_{k=1}^{n}\left(1+\frac{1}{k}\right) = \prod_{k=1}^{n}\left(\frac{k+1}{k}\right) = \frac{2}{1}\cdot\frac{3}{2}\cdot\frac{4}{3}\cdots\frac{n+1}{n} = n+1 \longrightarrow \infty,$$

we have another proof of the divergence of the harmonic series.

Hyperbolic Functions

The series which defines the exponential function

$$e^x = 1 + x + \frac{x^2}{2!} + \frac{x^3}{3!} + \frac{x^4}{4!} + \frac{x^5}{5!} + \cdots + \frac{x^k}{k!} + \cdots$$

converges absolutely for any real number x. Then if we add only some of its terms, we will have a series which also converges absolutely for any real number x.

We may then define the "even part" and the "odd part" of e^x by summing only the even-indexed terms or only the odd-indexed terms:

$$\begin{aligned}
\cosh x &= 1 + \frac{x^2}{2!} + \frac{x^4}{4!} + \frac{x^6}{6!} + \cdots + \frac{x^{2k}}{(2k)!} + \cdots \\
\sinh x &= x + \frac{x^3}{3!} + \frac{x^5}{5!} + \frac{x^7}{7!} + \cdots + \frac{x^{2k+1}}{(2k+1)!} + \cdots
\end{aligned}$$

By doing this, we obtain the functions *hyperbolic cosine* and *hyperbolic sine* (in the exercises you will see the reason for these names). Clearly the hyperbolic cosine is an *even* function, while the hyperbolic sine is an *odd* function in the following sense:

$$\begin{aligned}
\cosh(-x) &= \cosh x \\
\sinh(-x) &= -\sinh x.
\end{aligned}$$

Adding both series, we obtain $e^x = \cosh x + \sinh x$, and bearing in mind the comment we just made regarding their parity, we have $e^{-x} = \cosh x - \sinh x$. From this we immediately obtain another way of writing the hyperbolic functions:

$$\cosh x = \frac{e^x + e^{-x}}{2}, \text{ and}$$

$$\sinh x = \frac{e^x - e^{-x}}{2}.$$

The *hyperbolic tangent* may be defined by $\tanh x = \frac{\sinh x}{\cosh x} = \frac{e^x - e^{-x}}{e^x + e^{-x}}$.

Injectivity and Inverse Functions

Before introducing other functions we need to consider two basic ideas linked to the abstract notion of function: domain and injectivity.

We have said that a function f assigns to each real number x another real number $f(x)$. Often, either for convenience or by necessity, we will not permit the variable x to take all real values but only some, restricting the values of x to some subset $A \subset \mathbb{R}$. We will call this subset the *domain* of f.

Thus, for example, if we consider the problem: "of all rectangles with perimeter four, which has the largest area?", we will seek the value of x maximizing

$$\text{area} = \text{base} \times \text{height} = x(2 - x) = -x^2 + 2x = f(x),$$

but as neither the base x nor the height $2 - x$ make sense for this problem if they are negative, we consider only $x > 0$ and $2 - x > 0$. In other words, we put

$$f : (0, 2) \longrightarrow \mathbb{R} \text{ such that } f(x) = -x^2 + 2x.$$

$(0, 2)$ is the domain of f. Another example: we want to study the function $g(x) = \frac{1}{x}$, but since this function is not defined for the value $x = 0$, we consider $A = \{x \in \mathbb{R} : x \neq 0\}$, and this is the domain of the function we wish to study:

$$g : A \longrightarrow \mathbb{R} \text{ such that } g(x) = \frac{1}{x}.$$

Another: $\tan x = \frac{\sin x}{\cos x}$ is defined only when $\cos x \neq 0$, so we will have to consider a domain which does not contain the points $\frac{\pi}{2} + k\pi$ with $k \in \mathbb{Z}$.

Many times we will be interested in—given the value $f(x)$—recovering x. For example, for the function $g(x) = \frac{1}{x}$ above, if $g(x) = \frac{3}{2}$, what is the value of x? Set

$$\frac{1}{x} = \frac{3}{2},$$

from where we "solve for x": $x = \frac{2}{3}$. But it will sometimes happen that it is impossible to determine x: in the previous example ($f(x) = -x^2 + 2x$), if we know that $f(x) = \frac{3}{4}$, what is the value of x?

$$\frac{3}{4} = -x^2 + 2x$$
$$0 = x^2 - 2x + \frac{3}{4}.$$

Solving for x, we have either $x = \frac{1}{2}$ or $x = \frac{3}{2}$. What happens is that the function f assigns the value $\frac{3}{4}$ to two different values of x: $f(\frac{1}{2}) = \frac{3}{4} = f(\frac{3}{2})$, so it is not possible to determine x.

We will say that a function is *injective* or *one-to-one* when each $f(x)$ comes from one unique value of the variable x. Another way of saying this is that $x \neq y$ implies $f(x) \neq f(y)$ (in other words, $f(x) = f(y)$ implies $x = y$). When this happens, given $f(x)$, it will be possible to solve for x and "recover" its value.

For example, the exponential function $f(x) = e^x$ is injective: if $x \neq y$, suppose $x < y$ and write $y = x + h$ with $h > 0$. Then

$$e^y = e^{x+h} = e^x e^h > e^x, \text{ for } e^h = \sum_{k=0}^{\infty} \frac{h^k}{k!} = 1 + \sum_{k=1}^{\infty} \frac{h^k}{k!} > 1.$$

This argument actually shows something stronger: the exponential function is strictly increasing: $x < y$ implies $e^x < e^y$.

When a function is not injective in all its domain, we can often consider a smaller domain in which it is injective. The function

$$f : \mathbb{R} \longrightarrow \mathbb{R} \text{ such that } f(x) = x^2$$

is not injective, for $f(x) = f(-x)$ for any x. But if we consider as its domain $A = [0, \infty)$,

$$f : A \longrightarrow \mathbb{R} \text{ such that } f(x) = x^2$$

is injective. Likewise, the functions $\cos x$ and $\sin x$ are not injective, but restricting their domains

$$\cos : [0, \pi] \longrightarrow \mathbb{R}$$
$$\sin : \left[-\frac{\pi}{2}, \frac{\pi}{2}\right] \longrightarrow \mathbb{R}$$

they are.

The following functions—with the domains given—are injective. I've also written where they take values:

$$
\begin{aligned}
x &\mapsto x^2, & [0,\infty) &\longrightarrow [0,\infty),\\
x &\mapsto e^x, & \mathbb{R} &\longrightarrow (0,\infty),\\
x &\mapsto \cos x, & [0,\pi] &\longrightarrow [-1,1],\\
x &\mapsto \sin x, & \left[-\frac{\pi}{2},\frac{\pi}{2}\right] &\longrightarrow [-1,1],\\
x &\mapsto \tan x, & \left(-\frac{\pi}{2},\frac{\pi}{2}\right) &\longrightarrow \mathbb{R}.
\end{aligned}
$$

Because of their injectivity, in each case it is possible to recover x given $y = f(x)$. Thus, for example, if $y = x^2$, $x = \sqrt{y}$ (understanding that this means the positive root of x^2). It is possible to define in each case, the *inverse function* $x = f^{-1}(y)$ which recovers x, given y:

$$
\begin{aligned}
y &\mapsto \sqrt{y}, & [0,\infty) &\longrightarrow [0,\infty), \text{ the square root},\\
y &\mapsto \ln y, & (0,\infty) &\longrightarrow \mathbb{R}, \text{ the natural logarithm},\\
y &\mapsto \arccos y, & [-1,1] &\longrightarrow [0,\pi], \text{ the arc cosine},\\
y &\mapsto \arcsin y, & [-1,1] &\longrightarrow \left[-\frac{\pi}{2},\frac{\pi}{2}\right], \text{ the arc sine},\\
y &\mapsto \arctan y, & \mathbb{R} &\longrightarrow \left(-\frac{\pi}{2},\frac{\pi}{2}\right), \text{ the arc tangent}.
\end{aligned}
$$

Note that we do not necessarily have a formula to calculate the values of these functions. But we *do* know that they exist. For example, we know, because of the injectivity of $f(x) = e^x$, that there is a *unique* value of x for which $e^x = y$. We call this value $x = \ln y$ even though we have—as yet—no way of calculating it.

When the values of a function fall within the domain of another, it is possible to "compose" them, in the following sense:

$$
A \xrightarrow{f} B \xrightarrow{g} C
$$

$$
x \mapsto f(x) \mapsto g(f(x)).
$$

The function $x \mapsto g(f(x))$ is called *g composed with f* and denoted $g \circ f$. For example, if $f(x) = x^2$ and $g(x) = e^x$, we have $(g \circ f)(x) = e^{x^2}$. Note that this is not the same as $(f \circ g)(x) = (e^x)^2 = e^{2x}$. If f has an inverse f^{-1}, composing one with the other we obtain the "identity" function: $x \mapsto x$. For example,

$$
\arcsin(\sin x) = x \quad \text{and} \quad e^{\ln y} = y.
$$

Curves in the Plane: Parametrized Curves

We need to consider curves in the plane. We will use the following three ways to describe a curve $\mathcal{C} \subset \mathbb{R}^2$.

(a) The *parametric form*: $\mathcal{C}$ is the image of a function

$$p : \mathbb{R} \longrightarrow \mathbb{R}^2,$$

where $p(t)$ "draws" the curve $\mathcal{C}$ as the "parameter" t moves.
(b) The *implicit form*: $\mathcal{C}$ is the set of points of the plane on which a function

$$F : \mathbb{R}^2 \longrightarrow \mathbb{R}$$

is zero.
(c) As the *graph of a function*: $\mathcal{C} = \{(x, f(x)) : x \in \mathbb{R}\}$, where

$$f : \mathbb{R} \longrightarrow \mathbb{R}.$$

Example But let's see one particular example represented in each of these three forms: the circumference $\mathcal{C}$ centered at $(0, 0)$ and of radius 1.

(a) In parametric form: $\mathcal{C}$ is the image of $p : [0, 2\pi] \longrightarrow \mathbb{R}^2$ where:

$$p(t) = (\cos t, \sin t).$$

In this representation the point $p(t)$ moves along the circumference anti-clockwise, from the point $(1, 0)$ and does one full turn as t moves from 0 to 2π. One could parametrize the circumference in many other ways; we will see another way below.
(b) In implicit form: the circumference $\mathcal{C}$ is formed by the points (x, y) of the plane whose distance from $(0, 0)$ is one. We may write

$$1 = \text{ distance from } (0, 0) \text{ to } (x, y) = \sqrt{x^2 + y^2},$$

so these points are those which verify the equation $x^2 + y^2 = 1$. Thus the circumference is the set of points (x, y) where the function

$$F(x, y) = x^2 + y^2 - 1$$

is zero.

(c) As the graph of a function: the upper half of the circumference $\mathcal{C}$ may be described as the set of points $(x, f(x))$, where $f : [-1, 1] \longrightarrow \mathbb{R}$ is

$$f(x) = \sqrt{1 - x^2}.$$

The lower half is the graph of $g(x) = -\sqrt{1 - x^2}$.

As we see in this example, one of the limitations of this form of representation is that there are many curves which are not graphs of functions. All three ways of representing curves $\mathcal{C} \subset \mathbb{R}^2$ are useful and we will use, in any given case, that which is most convenient.

The Cycloid

We consider now another example of curve, which we will use later in Chap. 4: the Cycloid. Take a circumference of radius r centered at $(0, -r)$. The point A of the circumference is resting on the x-axis coinciding with $(0, 0)$. But now we begin to roll the circumference along the x-axis, without sliding, and we mark the curve described by point A. This is a cycloid. Let's parametrize it.

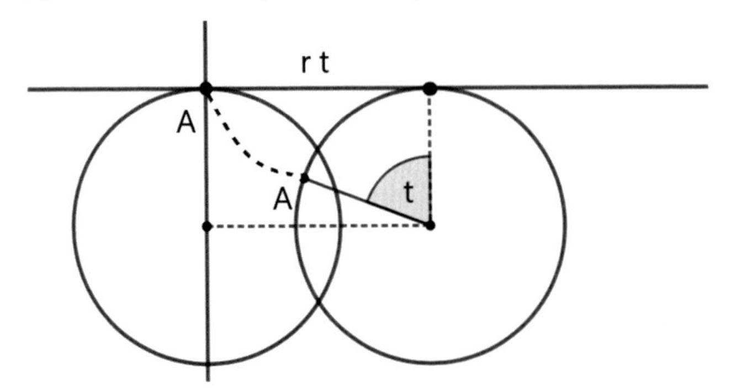

Say that the circumference has turned t radians. This turn comprises an arc of length rt, so the circumference has moved to the right by this amount. Where is point A now? It has moved to the right $rt - r \sin t$; and its height is now $-r + r \cos t$. Its new position is therefore $(rt - r \sin t, -r + r \cos t)$.

Thus the cycloid may be parametrized by $p : [0, 2\pi] \to \mathbb{R}^2$:

$$p(t) = r(t - \sin t, -1 + \cos t).$$

Finally, the cycloid looks like this

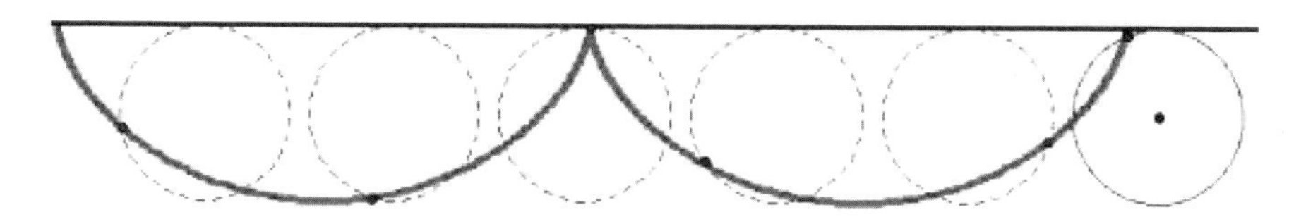

(after two turns, that is $t \in [0, 4\pi]$).

Pythagorean Triples

We will use a "rational" parametrization of the circumference to obtain all Pythagorean triples. Recall that the Pythagorean Theorem states that given a right triangle, the area of the square whose side is the hypotenuse is equal to the sum of the areas of the squares on the other two sides (the legs). This holds for any right triangle; the lengths of its sides can be natural, rational, or irrational numbers. However, the usual example is the triangle whose sides measure 3, 4, and 5. A triple such as this one, in which the lengths of the hypotenuse and both legs of the right triangle are all natural numbers, is called a Pythagorean triple. If we have natural numbers a, b, and c such that $a^2 + b^2 = c^2$, dividing by c^2 we have

$$\left(\frac{a}{c}\right)^2 + \left(\frac{b}{c}\right)^2 = 1,$$

thus, having a Pythagorean triple is like having a "rational point" of the circumference. That is, a point on the circumference whose coordinates are both rational numbers. We will define a parametrization $p : [0, 1] \longrightarrow \mathbb{R}^2$ of the circumference (actually of the upper right-hand quarter of the circumference) which has the property of making the rational numbers t on the interval $[0, 1]$ correspond to the rational points $p(t)$ of the circumference. To do this let's begin by looking at the following drawing:

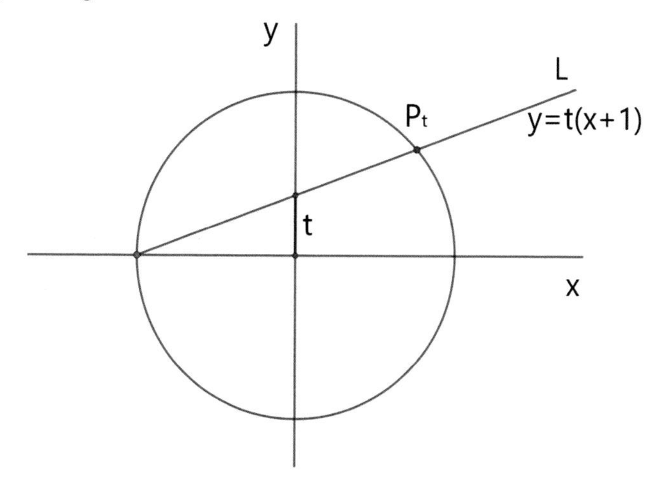

We want to make t correspond to P_t, on the intersection of the circumference and the line $y = t(x+1)$ (which goes through $(-1, 0)$ and $(0, t)$, as in the drawing). Let's find that intersection:

$$\begin{aligned} x^2 + y^2 &= 1 \\ x^2 + (t(x+1))^2 &= 1 \\ x^2 + t^2(x^2 + 2x + 1) &= 1 \\ (t^2+1)x^2 + 2t^2x + (t^2-1) &= 0, \end{aligned}$$

from which (finding the roots of this quadratic equation), we obtain $x = -1$ (which corresponds to the point of intersection $(-1, 0)$) or

$$x = \frac{-2t^2 + \sqrt{4t^4 - 4(t^4-1)}}{2(t^2+1)} = \frac{-t^2+1}{t^2+1},$$

which corresponds to the point of intersection $(\frac{-t^2+1}{t^2+1}, \frac{2t}{t^2+1})$. We define then

$$p : [0, 1] \longrightarrow \mathbb{R}^2 \qquad \text{such that } p(t) = \left(\frac{-t^2+1}{t^2+1}, \frac{2t}{t^2+1}\right).$$

Note that if t is rational, so are the coordinates of the point $p(t)$. Now, if $P = (a, b)$ is a rational point of the circumference, looking at

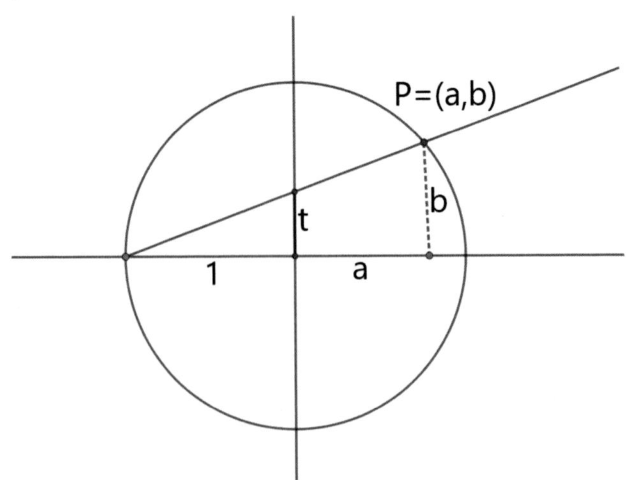

we see that, by triangle similarity, $t = \frac{t}{1} = \frac{b}{1+a}$, which is therefore rational. Thus the rational numbers of $[0, 1]$ correspond to the rational points of the first quadrant of the circumference. We have then, for any rational t

$$\left(\frac{-t^2+1}{t^2+1}\right)^2+\left(\frac{2t}{t^2+1}\right)^2=1,$$

then

$$(1-t^2)^2+(2t)^2=(1+t^2)^2.$$

Setting $t=\frac{m}{n}$, and multiplying by n^4

$$(n^2-m^2)^2+(2mn)^2=(n^2+m^2)^2,\ \text{with}\ m\leq n,$$

so n^2-m^2, $2mn$, n^2+m^2 is a Pythagorean triple for any pair of natural numbers $m<n$. We may thus describe *all* Pythagorean triples (up to multiples). For example, for $m=356$ and $n=921$, we obtain the triple

$$721505 \qquad 655752 \qquad 974977.$$

Continuity

In Chap. 1 we justified the need for completeness, and for passing from the rational line to the real line, by saying that we wanted there to actually be intersection in a situation such as

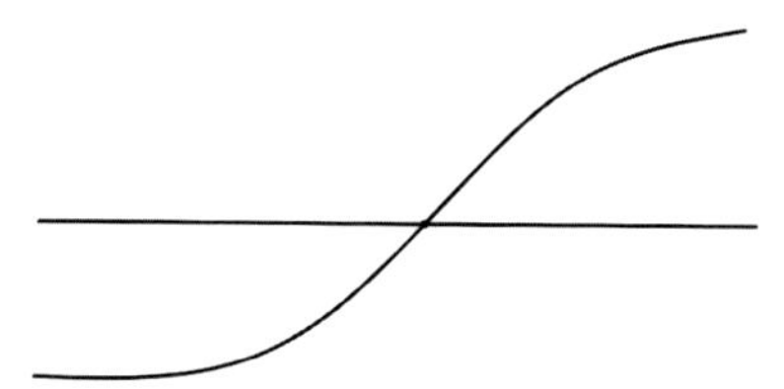

However, this drawing implicitly verifies a second condition which is essential for the curve not to jump from one side of the line to the other: continuity. A function is continuous when small changes of the variable x produce small changes of the values $f(x)$, in other words the values $f(x)$ do not jump around; they rather have a certain quality of permanence which we will formalize as follows.

We will say that a function f is *continuous at the point* s if

$$f(s)<v\Rightarrow f(x)<v \text{ for all } x \text{ in some neighborhood of } s$$

$$\text{and } f(s)>u\Rightarrow f(x)>u \text{ for all } x \text{ in some neighborhood of } s.$$

When f has this property at all points s in its domain, we simply say that *f is continuous*.

For example, the function $f(x) = 2x$ is continuous: if $u < 2s < v$, we obtain $u < 2x < v$ whenever $\frac{u}{2} < x < \frac{v}{2}$, which happens for x sufficiently close to s, for $\frac{u}{2} < s < \frac{v}{2}$.

The following Lemma will connect what we have seen on convergent sequences, with the properties of continuous functions.

Lemma *If* $x_n \to s$ *and* f *is continuous at* s*, then* $f(x_n) \to f(s)$.

To see why, let's start by taking any $\varepsilon > 0$. Since

$$f(s) - \varepsilon < f(s) < f(s) + \varepsilon,$$

the continuity of f at s assures us that for all x in some neighborhood of s—say $(s - \delta, s + \delta)$—we will also have

$$f(s) - \varepsilon < f(x) < f(s) + \varepsilon.$$

But, on the other hand, since $x_n \to s$, for n sufficiently large (say $n \geq n_0$) x_n are in the neighborhood $(s - \delta, s + \delta)$. Thus,

$$f(s) - \varepsilon < f(x_n) < f(s) + \varepsilon, \qquad \text{for all } n \geq n_0.$$

In other words, $|f(x_n) - f(s)| < \varepsilon$ for sufficiently large n. □

Bolzano and Weierstrass

Bernardus Placidus Johann Gonzal Nepomuk (!) Bolzano (1781–1848) was born in Prague. Mathematician and philosopher, he gave the first formal definition of limit and recognized in modern terms, the completeness property of the real numbers.

Continuity, together with the completeness of the real line, will ensure the existence of intersections, according to the following theorem.

Bolzano's Theorem *if* $f : [a, b] \longrightarrow \mathbb{R}$ *is continuous,* $f(a) < 0$ *and* $f(b) > 0$*, then for some* $s \in (a, b)$*,* $f(s) = 0$.

Let's see the proof: consider the set

$$A = \{x \in [a, b] : f(x) < 0\}.$$

A is non-empty ($a \in A$) and bounded above (by b).

The completeness axiom therefore assures the existence in $[a, b]$, of the supremum of A: call it s. We will see that $f(s) = 0$:

If $f(s) < 0$, since f is continuous at s the same would hold for any x close to s (that is, $f(x) < 0$ for all $x \in (s - \varepsilon, s + \varepsilon)$). But then $(s - \varepsilon, s + \varepsilon) \subset A$ and we would have $s < x \in A$: s would not be an upper bound of A.

If $f(s) > 0$, since f is continuous at s the same would hold for any x close to s (that is, $f(x) > 0$ for all $x \in (s - \varepsilon, s + \varepsilon)$). But then we would have elements x, bounds of A, smaller than s: then s would not be the smallest upper bound of A.

Only one possibility is left: $f(s) = 0$. □

The following theorem is also very important.

Weierstrass' Theorem *If $f : [a, b] \longrightarrow \mathbb{R}$ is continuous, f attains a maximum value in $[a, b]$ (that is: there exists a $y \in [a, b]$ such that $f(y) \geq f(x)$ for all $x \in [a, b]$).*

Let's see why: we will first see that the set of values attained by the function is bounded above; and then we will see that its supremum is one of the values attained.

First consider the set

$$A = \{f(x) : x \in [a, b]\}.$$

We will prove that A is bounded above. If it were not, there would exist for each $n \in \mathbb{N}$ an element $x_n \in [a, b]$ such that $n < f(x_n)$. As the sequence (x_n) is contained in the closed interval $[a, b]$, it has a subsequence (x_{n_k}) which converges to a point x in $[a, b]$. Since f is continuous, $f(x_{n_k}) \to f(x)$. But this cannot be, because for all k, $n_k < f(x_{n_k})$, so $f(x_{n_k})$ cannot converge to any real number.

So A is bounded above and non-empty. By completeness, it has a supremum $s = \sup A$. Take—for each n—$y_n \in [a, b]$ such that

$$s - \frac{1}{n} < f(y_n) \leq s.$$

Again, (y_n) has a convergent subsequence; say $y_{n_k} \to y \in [a, b]$. Then

$$\begin{array}{ccc} s - \dfrac{1}{n_k} & < f(y_{n_k}) & \leq s \\ \downarrow & \downarrow & \\ s & \leq f(y) & \leq s. \end{array}$$

Thus, $f(y) = s$ and the maximum s is attained on an element of $[a, b]$. □

An analogous argument proves that every continuous function on an closed interval also attains its minimum value.

Limits

We will say that *the limit of* $f(x)$ *as* x *tends to* s *is* L and we will write

$$\lim_{x \longrightarrow s} f(x) = L$$

if

Given any neighborhood V of L, $f(x) \in V$ for all x sufficiently close to s (but different to s).

Example $\lim_{x \longrightarrow 0} \cos x = 1$.

The following drawing shows that for any $\varepsilon > 0$, if x is small enough, $\cos x$ will be between $1 - \varepsilon$ and 1:

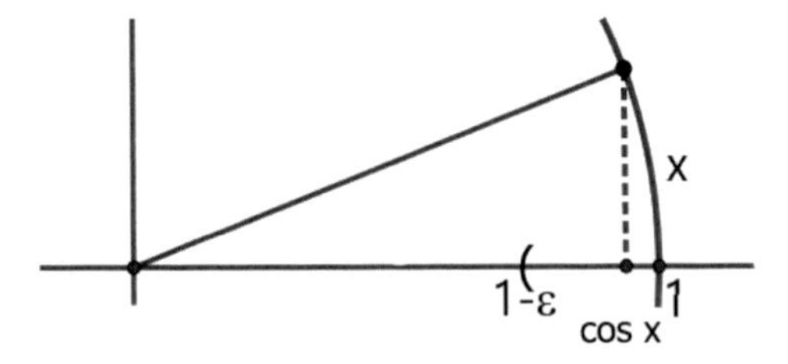

There is a way of expressing the continuity of f at s in terms of limits:

Proposition *f is continuous at s if and only if* $\lim_{x \longrightarrow s} f(x) = f(s)$.

Let's see why: Say that f is continuous at s and take $(f(s) - \varepsilon, f(s) + \varepsilon)$ a neighborhood of $f(s)$. Since $f(s) > f(s) - \varepsilon$, then also $f(x) > f(s) - \varepsilon$ for all x sufficiently close to s. And since $f(s) < f(s) + \varepsilon$, also $f(x) < f(s) + \varepsilon$ for all x sufficiently close to s. Thus $f(x) \in (f(s) - \varepsilon, f(s) + \varepsilon)$ for all x close to s.

On the other hand, if we know that $\lim_{x \longrightarrow s} f(x) = f(s)$, given $u < f(s)$, let ε be so small that $u < f(s) - \varepsilon$. But we know that $f(x) \in (f(s) - \varepsilon, f(s) + \varepsilon)$ for all x close to s. In particular, $u < f(x)$ for all x close to s. The same if $v > f(s)$. Thus, f is continuous at s, according to our definition. □

Limits in Ancient Greece: The Area of a Circle

The notion of limit, as we know it today, was formalized by Bolzano in 1817. It is, however, an ancient idea. Greek mathematicians such as Antiphon, Bryson, Eudoxus and Archimedes used arguments of the following type: suppose they want to prove that $\alpha = \beta$. They start by constructing two sequences of positive numbers A_n and B_n such that, for each n,

$$\begin{aligned} A_n &\leq \alpha \leq B_n \\ A_n &\leq \beta \leq B_n. \end{aligned} \tag{3.1}$$

Then if, given *any* number c larger than one, they could find an n for which $\frac{B_n}{A_n} < c$, they would conclude that $\alpha = \beta$.

The conclusion is correct, for if in (3.1) we divide by A_n, we have

$$\begin{aligned} 1 &\leq \frac{\alpha}{A_n} \leq \frac{B_n}{A_n} < c \\ 1 &\leq \frac{\beta}{A_n} \leq \frac{B_n}{A_n} < c, \end{aligned}$$

and then

$$\begin{aligned} |\alpha - \beta| &\leq |\alpha - A_n| + |A_n - \beta| \\ &= (\alpha - A_n) + (\beta - A_n) \\ &= A_n\left(\frac{\alpha}{A_n} - 1\right) + A_n\left(\frac{\beta}{A_n} - 1\right) \\ &< A_n(c-1) + A_n(c-1) \\ &< \alpha(c-1) + \beta(c-1) \\ &= (\alpha + \beta)(c-1), \end{aligned}$$

but $(c - 1)$ can be as small as required, so $\alpha = \beta$. Let's see a concrete example of this type of argument.

Example Area of a circle of radius r.

Consider a circle of radius r, and call A its area and C the length of its circumference. We will use the method described above to prove that $2A = rC$. We begin by inscribing and circumscribing n-sided regular polygons in our circle.

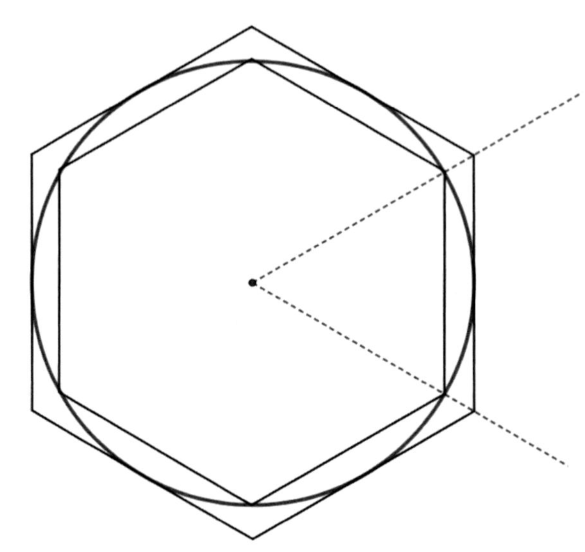

let's take a closer look at the following portion of the picture:

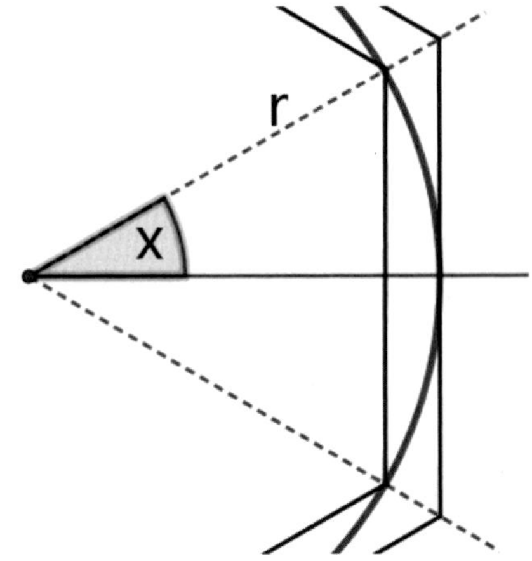

Comparing the areas of inscribed polygon, circle, and circumscribed polygon in this portion, we see that

$$r \cos x r \sin x < \frac{A}{n} < r^2 \tan x$$

and comparing the lengths of the boundaries, we have

$$2r \sin x < \frac{C}{n} < 2r \tan x.$$

Multiply the first by $2n$ and the second by rn, to obtain

$$2nr^2 \cos x \sin x < 2A < 2nr^2 \tan x$$
$$2nr^2 \sin x < rC < 2nr^2 \tan x.$$

Thus, both $2A$ and rC are between

$$A_n = 2nr^2 \cos x \sin x \quad \text{and} \quad B_n = 2nr^2 \tan x.$$

But

$$\frac{B_n}{A_n} = \frac{2nr^2 \tan x}{2nr^2 \cos x \sin x} = \frac{1}{\cos^2 x} \longrightarrow 1$$

as n grows, for the angle x will get smaller as we increase the number of sides n of our polygons. Then, for any $c > 1$ we will find n such that $\frac{B_n}{A_n} < c$, and we conclude that $2A = rC$.

Now, $C = 2\pi r$ (for this is how π is defined), so what we have proved is that $2A = r2\pi r = 2\pi r^2$, that is, the area of the circle of radius r is

$$A = \pi r^2.$$

I should note here that their argument would have been slightly different: sine and cosine were invented by Indian mathematicians of the Vth Century; the Greeks used the "chord" (as in the chord of a bow... chord$x = 2 \sin \frac{x}{2}$).

Note also that $\frac{A_n}{2}$ and $\frac{B_n}{2}$ are—respectively—the areas of the inscribed and circumscribed n-polygons, so for $r = 1$, we have $A_n \longrightarrow 2A = C = 2\pi$. Archimedes uses this fact to approximate π: he starts with $n = 6$ and doubles the number of sides of the polygons: 12, 24, 48, 96. His approximation of π was unsurpassed for centuries.

Three Important Limits

This is a good place to calculate some limits that we will need in the next chapter.

(i) $\lim_{x \longrightarrow 0} \frac{\sin x}{x} = 1$:
Look again at the (upper half of) the last figure, and consider $r = 1$,

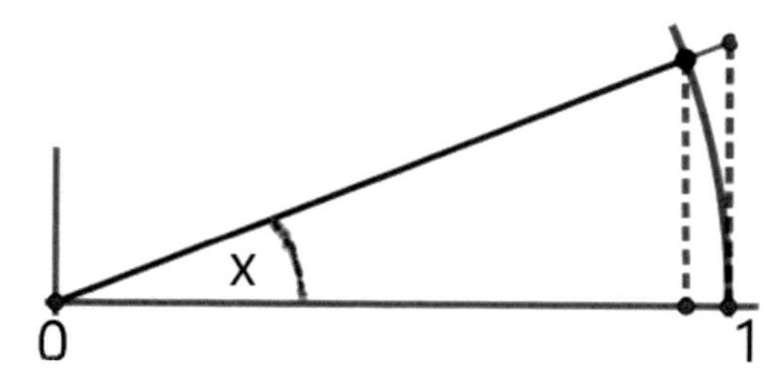

As the area of the smaller triangle is less than that of the portion of circle, which in turn is less than the area of the larger triangle, we have

$$\frac{\sin x \cos x}{2} < \frac{x}{2\pi}\pi < \frac{\tan x}{2}.$$

Multiplying by two,

$$\sin x \cos x < x < \frac{\sin x}{\cos x}.$$

From the first inequality, $\frac{\sin x}{x} < \frac{1}{\cos x}$. From the second, $\cos x < \frac{\sin x}{x}$. Thus,

$$\cos x < \frac{\sin x}{x} < \frac{1}{\cos x}.$$

Now, as $x \longrightarrow 0$, $\cos x \longrightarrow 1$, and we have

$$1 \leq \lim_{x \longrightarrow 0} \frac{\sin x}{x} \leq 1.$$

(ii) $\lim_{x \longrightarrow 0} \frac{\cos x - 1}{x} = 0$:
To see this, set

$$\frac{(\cos x - 1)(\cos x + 1)}{x} = \frac{\cos^2 x - 1}{x} = \frac{-\sin^2 x}{x} = \frac{\sin x}{x}(-\sin x).$$

Then,

$$\frac{\cos x - 1}{x} = \frac{\sin x}{x} \frac{(-\sin x)}{(\cos x + 1)},$$

which, if $x \longrightarrow 0$, tends to $1 \cdot \frac{0}{2} = 0$.

(iii) $\lim_{x \longrightarrow 0} \frac{e^x - 1}{x} = 1$:

$$\text{Write } \frac{e^x - 1}{x} = \frac{1}{x}\left(\sum_{k=0}^{\infty} \frac{x^k}{k!} - 1\right)$$

$$= \frac{1}{x} \sum_{k=1}^{\infty} \frac{x^k}{k!}$$

$$= \sum_{k=1}^{\infty} \frac{x^{k-1}}{k!}$$

$$= 1 + \sum_{k=2}^{\infty} \frac{x^{k-1}}{k!}$$

$$= 1 + x \sum_{k=2}^{\infty} \frac{x^{k-2}}{k!},$$

which, as $x \longrightarrow 0$ tends to $1 + 0 \cdot \textit{something}$ (note that the series converges absolutely by the quotient criterion).

Exercises

1 We have seen that if $x_n = \frac{1}{2}\frac{360^\circ}{n}$ is half of the angle containing one side of the inscribed n-sided regular polygon,

$$\frac{A_n}{2} = n \cos x_n \sin x_n \longrightarrow \pi.$$

Do as Archimedes: starting with the hexagon ($n = 6$), double repeatedly the number of sides to approximate π. Hint: use the half-angle formulas (which may be deduced from $\cos(x + x) = \cos x \cos x - \sin x \sin x$)

$$\cos\left(\frac{x}{2}\right) = \sqrt{\frac{1 + \cos x}{2}} \qquad \sin\left(\frac{x}{2}\right) = \sqrt{\frac{1 - \cos x}{2}}.$$

2 By using the properties of the exponential function, prove that

$$\ln(xy) = \ln x + \ln y,$$
$$\ln x^c = c \ln x.$$

3 Show that the functions $\cosh x$ and $\sinh x$ verify the following equality

$$(\cosh x)^2 - (\sinh x)^2 = 1.$$

Thus, the function $p : \mathbb{R} \longrightarrow \mathbb{R}^2$ given by $p(t) = (\cosh t, \sinh t)$ parametrizes a hyperbola (which justifies their name).

4 Find the image sets of the hyperbolic functions, and prove that their inverse functions are

$$\operatorname{arccosh} y = \ln\left(y + \sqrt{y^2 - 1}\right)$$

$$\operatorname{arcsinh} y = \ln\left(y + \sqrt{y^2 + 1}\right)$$

$$\operatorname{arctanh} y = \frac{1}{2}\ln\left(\frac{1+y}{1-y}\right).$$

5 The Fibonacci sequence (F_n) is defined by: $F_0 = 1$, $F_1 = 1$, and for all $n \geq 2$, $F_n = F_{n-2} + F_{n-1}$. Prove that if x, y, w, z are consecutive numbers of the Fibonacci sequence, then $(xz, 2yw, xz + 2y^2)$ is a Pythagorean triple.

6 f is called *increasing* if $x \leq y$ implies $f(x) \leq f(y)$, and *decreasing* if $x \leq y$ implies $f(x) \geq f(y)$. Which of the following are true? Prove those that are, and give a counterexample to those that are not.

(a) if f and g are increasing, then so is $f + g$.
(b) if f and g are increasing, then so is fg.
(c) if $f(x) \leq f(x+1)$ for all x, f is increasing.

7 Consider the function

$$f(x) = \begin{cases} x, & \text{if } x \in \mathbb{Q} \\ -x, & \text{if } x \notin \mathbb{Q}. \end{cases}$$

(i) Prove that it is neither increasing nor decreasing in any interval.
(ii) Find its inverse function f^{-1}.
(iii) Conclude that $g \circ f$ may be increasing though f and g are not.

8 Study the continuity of the function of Exercise 7.

9 Define a function that is continuous only at the point a. Define another that is continuous only at a and b.

10 Recall that the set of rational numbers is countable, and consider a numbering of $\mathbb{Q}$:

$$\mathbb{Q} = \{r_1, r_2, r_3, \ldots\}.$$

Now, for each $x \in \mathbb{R}$, set $S_x = \{n \in \mathbb{N} : r_n \leq x\}$, and define $f : \mathbb{R} \longrightarrow \mathbb{R}$ by

$$f(x) = \sum_{n \in S_x} \frac{1}{2^n},$$

where we are adding over all indices $n \in S_x$. Prove that

(i) the function f is strictly increasing ($x < y$ implies $f(x) < f(y)$).
(ii) f is discontinuous at each point $a \in \mathbb{Q}$.
(iii) f is continuous at each point $b \notin \mathbb{Q}$.

11 Prove that if f and g are continuous at s, then $f + g$ is continuous at s.

12 Let $f : [a, b] \longrightarrow [c, d]$ be increasing and bijective (injective and onto). Prove that:

(i) its inverse f^{-1} es increasing, and
(ii) f is continuous.

13 Let f and g be continuous on the interval $[a, b]$. If $f(a) > g(a)$ and $f(b) < g(b)$, prove that there is a $c \in (a, b)$ such that $f(c) = g(c)$.

14 Prove that there are infinitely many values of x for which $x \cos x = 1$.

15 Consider if Bolzano's and Weierstrass' theorems hold for the following functions, and why.

$$f : [-1, 1] \to \mathbb{R} \text{ given by } f(x) = \begin{cases} \frac{1}{x}, & \text{if } x \neq 0 \\ 1, & \text{if } x = 0 \end{cases}$$

and

$$g : (-1, 1) \to \mathbb{R} \text{ given by } g(x) = x.$$

16 Suppose f is defined and bounded close to a, and for each $n \in \mathbb{N}$ set

$$m_n = \inf\left\{f(x) : x \in \left(a - \frac{1}{n}, a + \frac{1}{n}\right)\right\}$$

$$M_n = \sup\left\{f(x) : x \in \left(a - \frac{1}{n}, a + \frac{1}{n}\right)\right\}.$$

Show that

(i) $m_n \leq M_n$ for all n,
(ii) (m_n) is increasing and (M_n) is decreasing.

Conclude that, since f is bounded near a, (m_n) and (M_n) converge. Call

$$\liminf_{x \longrightarrow a} f(x) = \lim_{n \longrightarrow \infty} m_n,$$

$$\limsup_{x \longrightarrow a} f(x) = \lim_{n \longrightarrow \infty} M_n.$$

Show that if $\liminf_{x \longrightarrow a} f(x) = L = \limsup_{x \longrightarrow a} f(x)$, then

$$\lim_{x \longrightarrow a} f(x) = L.$$

The Derivative

4

In this chapter we introduce the derivative, and its geometric counterpart, the tangent line. After seeing some of its properties, we present the Mean Value Theorems and some of their consequences.

Derivative

Pierre de Fermat (1607–1665) was a lawyer and worked on his mathematics in his free time. He gave impulse, as did Descartes, to analytic geometry. His work on maxima and minima of functions and on tangents foreshadow the notion of derivative, which he used in particular cases. Newton recognizes that his first ideas on calculus came from Fermat's tangent method. He also contributed to number theory. He enunciated the famous "Fermat's Last Theorem" ($x^n + y^n = z^n$ does not admit whole number solutions for $n \geq 3$), which was proved 357 years later by Andrew Wiles.

The idea of derivative appears in the XVIIth Century in the work of Pierre de Fermat and others, in general in relation to questions like: How do the values of a function f vary close to a given point? The central problem is then, to study *change* near a given value of the variable.

Consider the case of a moving body. Say that $f(t)$ is the distance traveled by that body from a certain initial moment until time t. If we wish to know the distance covered between the instant $t = t_0$ and the instant $t = t_1$ this will be $f(t_1) - f(t_0)$. And if what we are interested in is the mean speed with which the body has moved, it will be $speed = \frac{distance}{time}$

$$\frac{f(t_1) - f(t_0)}{t_1 - t_0},$$

where we have divided the distance covered by the elapsed time.

I. Zalduendo, *Calculus off the Beaten Path*, SUMS Readings,
https://doi.org/10.1007/978-3-031-15765-3_4

If we wish to know the mean speed between the instant $t = t_0$ and another moment a little later, say $t_1 = t_0 + h$ (for a "small" h), we will calculate:

$$\frac{f(t_0 + h) - f(t_0)}{h},$$

and if we set smaller and smaller values of h,

$$\lim_{h \to 0} \frac{f(t_0 + h) - f(t_0)}{h}$$

will approximate something like "the speed at the moment $t = t_0$."

Tangents

Another type of problem related with the question of how and how much the values of a function f vary are the problems of tangents. Consider the following graph of a function f

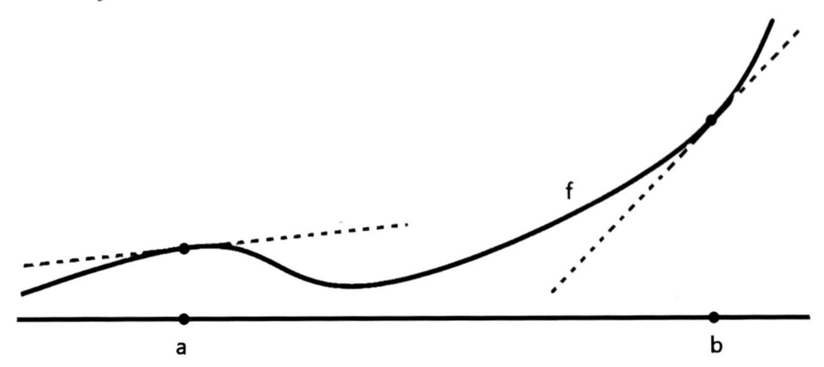

Clearly the function "changes little" close to the point a and "changes very much" close to b. And this is reflected by the slope of the lines tangent to the graph of f at those points: smaller slope = "little change," larger slope = "larger change." Having the line tangent to the graph of f at the point $(a, f(a))$ would simplify our question regarding the velocity of change in f close to the point a. Why? Because the tangent line will be the graph of a linear function such as $y = px+q$ (a polynomial of degree 1) whose growth depends solely on its "slope" p.

The following two functions coincide at $a = 1$. Imagine that you want to know which function grows faster near that point,

$$x^3 + 6x \quad \text{or} \quad 4x^2 + 2x + 1.$$

This is not evident... However, if you ask yourself which grows faster close to $a = 1$:

$$9x - 2 \quad \text{or} \quad 10x - 3$$

clearly it is the second, for it has a larger slope. These linear functions correspond to the tangents of the previous functions at $a = 1$, thus it is the second $(4x^2 + 2x + 1)$ which is growing faster at $a = 1$. One of the characteristics of differential calculus is precisely its capacity to simplify problems by "linearizing" functions.

Thus we will try to define the line tangent to the graph of a function f at the point $(a, f(a))$. This is the line which "best fits" the graph of f at that point:

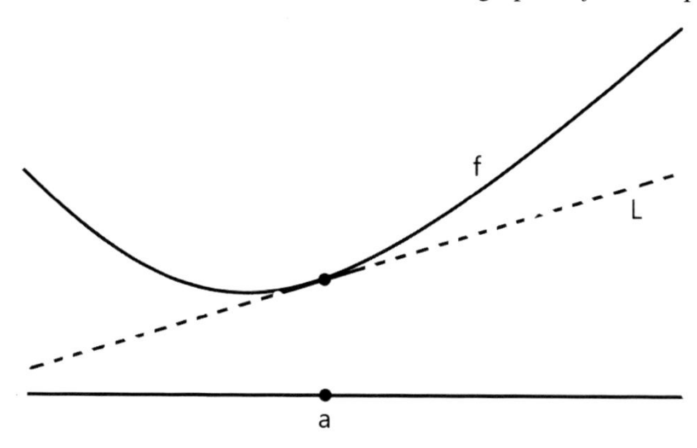

That is, it is the line L such that:

(i) goes through the point $(a, f(a))$, and which
(ii) has, at that point, the same "slope" as f.

A line is determined by two points on it, or also by its slope and one point on it. So conditions i) and ii) determine the line that we are looking for... the problem here is that we do not know the "slope." So what we will do is look at the slopes of lines which are close to the one we are looking for, and hope that their slopes will approximate the slope we want. Consider then the line L_h which goes through point $(a, f(a))$, and also $(a + h, f(a + h))$:

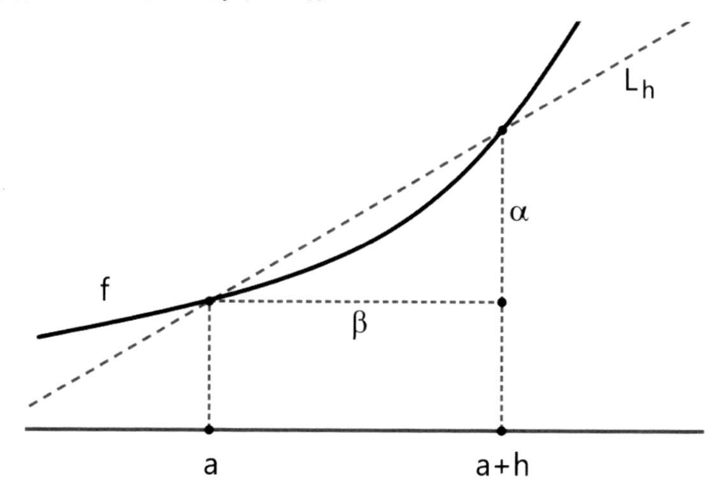

The slope of this line is $\frac{\alpha}{\beta}$, that is,

$$\frac{f(a+h) - f(a)}{h}.$$

Now, if we take smaller and smaller h, $a + h$ will approximate a, and hopefully the line L_h will approximate L. Thus we may consider that the slope of L is

$$\lim_{h \to 0} \frac{f(a+h) - f(a)}{h}.$$

Note the similarity with "speed" defined above. The slope of the tangent L is the speed of the change in f. We will then propose the following definition:

The *derivative of* f *at* a is the number

$$f'(a) = \lim_{h \to 0} \frac{f(a+h) - f(a)}{h}$$

(or equivalently: $f'(a) = \lim_{x \to a} \frac{f(x) - f(a)}{x - a}$).

Several comments are in order:

(i) The existence of this limit is not assured. In the exercises you will see an example where it does not exist. When it does, we will say that f is *differentiable at* a.

(ii) When f is differentiable at a, the tangent line L exists and has slope $f'(a)$. This line (again, see exercises) has an equation that can be expressed in any of the following ways:

$$y = f'(a)x + (f(a) - f'(a) \cdot a)$$
$$y = f'(a)(x - a) + f(a)$$
$$y = f(a) + f'(a)(x - a).$$

(iii) Note that when f is differentiable at a, in the quotient (called "difference quotient"):

$$\frac{f(a+h) - f(a)}{h}$$

both the numerator and the denominator, considered separately, tend to zero with h. This always happens and says absolutely nothing about the value of $f'(a)$.

(iv) Looking at the definition of $f'(a)$, we see that this number depends on the value of the function at a and at points x which are "very close" to a.

Nothing occurring at a (fixed) positive distance from a has any relevance to the derivative of f at a.

(v) When f is differentiable at x for all x in its domain, we will say simply that *f is differentiable*. In this case, a function sending each x to the value $f'(x)$ is defined. We denote it by f', and call this function *the derivative* of f.

Finally, another important property:

Proposition *When f is differentiable at a, it is necessarily continuous at a.*

To see this, write:

$$f(x) = f(a) + \frac{f(x) - f(a)}{x - a}(x - a)$$

and note that if $x \to a$, we have:

$$\lim_{x \to a} f(x) = f(a) + \lim_{x \to a} \frac{f(x) - f(a)}{x - a} \lim_{x \to a} (x - a)$$

$$\lim_{x \to a} f(x) = f(a) + f'(a) \cdot 0 = f(a)$$

which says precisely—as we have seen in Chap. 3—that f is continuous at a. Note that if f were not differentiable at a, $\lim_{x \to a} \frac{f(x)-f(a)}{x-a}$ could be infinite or non-existent, and the equality would not hold. □

Newton–Raphson

Let's see an interesting application of the idea of tangent line: The Newton–Raphson method for finding a zero of a function f. Say $f : \mathbb{R} \longrightarrow \mathbb{R}$ is differentiable, and we want to find point a where $f(a) = 0$. Observe the graph below

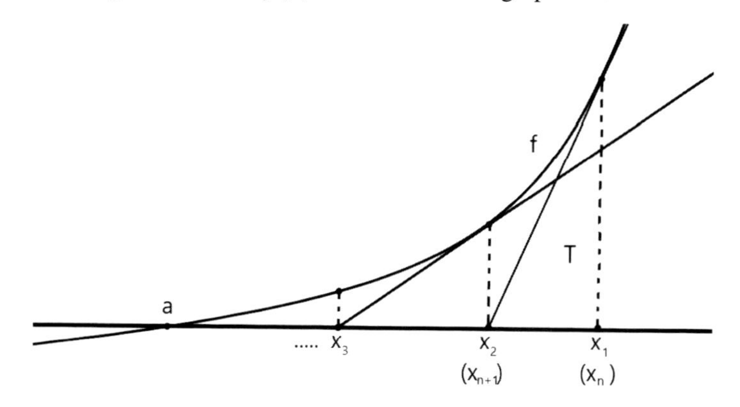

Take any point x_1, and consider the tangent line to the graph of f at point $(x_1, f(x_1))$. Now call x_2 the point of intersection of this line with the x-axis. Repeat the process to obtain x_3. Then, x_4, etc. At least in this drawing, the points $x_1, x_2, x_3, \ldots$ seem to converge to the point a where f is zero. Let's see how to pass from x_n to x_{n+1}: observing the triangle T we see that

$$f'(x_n) = \frac{f(x_n)}{x_n - x_{n+1}}.$$

From here we can solve for x_{n+1} in terms of x_n:

$$x_n - x_{n+1} = \frac{f(x_n)}{f'(x_n)}$$
$$x_n - \frac{f(x_n)}{f'(x_n)} = x_{n+1},$$

and this gives us an algorithm to construct the sequence (x_n). We see in the first line that, if $x_n \longrightarrow a$ (and $f'(a) \neq 0$),

$$0 = a - a = \frac{f(a)}{f'(a)}, \qquad \text{and then } f(a) = 0.$$

So if (x_n) converges, it converges to a point where f vanishes. For the sequence (x_n) to converge, it must happen that

$$|x_n - x_{n+1}| = \frac{|f(x_n)|}{|f'(x_n)|}$$

becomes smaller and smaller... This doesn't always happen; we will later see a condition that assures the convergence of the sequence (x_n). For now, let's see an example.

Example Approximation of the square root of c.

We wish to find the point $a = \sqrt{c}$, that is, a zero of the function $f(x) = x^2 - c$. Let's use the Newton–Raphson method:

$$\begin{aligned} x_{n+1} &= x_n - \frac{x_n^2 - c}{2x_n} \\ &= \frac{2x_n^2 - x_n^2 + c}{2x_n} \\ &= \frac{x_n^2 + c}{2x_n}, \end{aligned}$$

we construct the sequence $x_1, x_2, x_3, \ldots$, and we will see that $x_n \longrightarrow \sqrt{c}$. In other words, we have a method for approximating $\sqrt{c}$. This is usually called the *Babylonian method*, for this is how they did it... without knowing about tangents, or Newton, or Raphson. If you approximate $\sqrt{10}$ by this method starting with $x_1 = 3$, in a couple of steps you will have $3, 1623\ldots$, which is within one ten-thousandths of the true value.

Derivatives of the Elementary Functions

We will calculate the derivatives of some elementary functions. We leave several others as exercises at the end of the chapter.

The Derivative of $f(x) = x^n$: $f'(x) = nx^{n-1}$:
Take the difference quotient

$$\begin{aligned}
\frac{(x+h)^n - x^n}{h} &= \frac{1}{h}\left[\sum_{k=0}^{n}\binom{n}{k}x^{n-k}h^k - x^n\right] \\
&= \frac{1}{h}\left[\sum_{k=1}^{n}\binom{n}{k}x^{n-k}h^k\right] \\
&= \frac{1}{h}\left[h\sum_{k=1}^{n}\binom{n}{k}x^{n-k}h^{k-1}\right] \\
&= nx^{n-1} + \sum_{k=2}^{n}\binom{n}{k}x^{n-k}h^{k-1} \\
&= nx^{n-1} + h\sum_{k=2}^{n}\binom{n}{k}x^{n-k}h^{k-2}
\end{aligned}$$

which tends to nx^{n-1} as $h \to 0$.

Derivative of the Sine Function If $f(x) = \sin(x)$, $f'(x) = \cos(x)$.
Take the difference quotient

$$\begin{aligned}
\frac{\sin(x+h) - \sin(x)}{h} &= \frac{\sin(x)\cdot\cos(h) + \cos(x)\cdot\sin(h) - \sin(x)}{h} \\
&= \sin(x)\cdot\left(\frac{\cos(h) - 1}{h}\right) + \cos(x)\cdot\left(\frac{\sin(h)}{h}\right)
\end{aligned}$$

which, when $h \to 0$, converges to $\sin(x) \cdot 0 + \cos(x) \cdot 1 = \cos(x)$, bearing in mind the limits seen in Chap. 3.

Derivative of the Cosine Function If $f(x) = \cos(x)$, $f'(x) = -\sin(x)$.

Take the difference quotient

$$\frac{\cos(x+h) - \cos(x)}{h} = \frac{\cos(x) \cdot \cos(h) - \sin(x) \cdot \sin(h) - \cos(x)}{h}$$
$$= \cos(x) \cdot \left(\frac{\cos(h) - 1}{h}\right) - \sin(x) \cdot \left(\frac{\sin(h)}{h}\right)$$

which tends to $\cos(x) \cdot 0 - \sin(x) \cdot 1 = -\sin(x)$ as $h \to 0$.

Derivative of the Exponential Function If $f(x) = e^x$, $f'(x) = e^x$ (e^x is its own derivative!).

Take the difference quotient

$$\frac{e^{x+h} - e^x}{h} = \frac{e^x e^h - e^x}{h}$$
$$= e^x \left[\frac{e^h - 1}{h}\right]$$

which, when $h \to 0$, tends to e^x (for $\frac{e^h - 1}{h} \to 1$).

We also leave as an exercise the following.

Proposition *If f and g are differentiable, then so are their sum, product and quotient, and*

(i) $(f + g)'(x) = f'(x) + g'(x)$,
(ii) $(fg)'(x) = f'(x) \cdot g(x) + f(x) \cdot g'(x)$,
(iii) $\left(\frac{f}{g}\right)'(x) = \frac{f'(x) \cdot g(x) - f(x) g'(x)}{g(x)^2}$. □

With (iii), we can now differentiate the function $\tan x = \frac{\sin(x)}{\cos(x)}$:

$$(\tan x)' = \frac{\cos(x)\cos(x) - (\sin(x))(-\sin(x))}{\cos^2(x)} = \frac{\cos^2(x) + \sin^2(x)}{\cos^2(x)} = \frac{1}{\cos^2(x)}.$$

The Chain Rule

We now return to another operation that can be performed with functions: composition. Recall that we have defined

$$(g \circ f)(x) = g(f(x)).$$

We will see how to differentiate a composition of functions. The resulting formula, known as "the chain rule," is very important both for the theory and in applications. Its theoretical relevance will become apparent immediately, when we use it to calculate the derivatives of some inverse functions (logarithm, arcsine, arccosine,...) and in Chap. 5, when we see how to integrate by "changing variables" (also known as substitution).

Its importance in applications stems from the fact that it is often convenient, when a function can be considered as depending on different variables, to understand the rate of change of the function when varying one or the other. An example: a balloon is inflated by injecting air into it at a constant rate. We want to know how the radius of the balloon changes over time. The radius of the balloon may be considered as a function of its volume. We have: $r = r(v)$ and $v = v(t)$ (where r is the radius, v the volume, and t is time). So,

$$r = r(v(t))$$

in other words, the radius is—through composition—a function of time. The chain rule will tell us how the speed at which these variables change are linked.

The Chain Rule *If f is a function differentiable at x and g is a function differentiable at $f(x)$, then $g \circ f$ is differentiable at x and*

$$(g \circ f)'(x) = g'(f(x)) \cdot f'(x).$$

Let's see why. Write $k = f(x+h) - f(x)$. And note two things. The first: since f is continuous at x, as $h \to 0$ then also $k \to 0$. The second: $f(x+h) = f(x) + k$. Now let's consider the difference quotient:

$$\begin{aligned}\frac{(g \circ f)(x+h) - (g \circ f)(x)}{h} &= \frac{g(f(x+h)) - g(f(x))}{h} \\ &= \frac{g(f(x)+k) - g(f(x))}{k} \cdot \frac{k}{h} \\ &= \frac{g(f(x)+k) - g(f(x))}{k} \cdot \frac{f(x+h) - f(x)}{h}\end{aligned}$$

which, as $h \to 0$ (and therefore $k \to 0$) tends to $g'(f(x)).f'(x)$. □

Coming back to the example of the balloon, $r = r(v)$, but through composition, $r = r(v(t))$, the chain rule tells us

$$(r \circ v)'(t) = r'(v(t))\, v'(t).$$

Sometimes this is written using the notation

$$\frac{dr}{dt} = \frac{dr}{dv}\frac{dv}{dt},$$

which might be a valid mnemonic device, but we should not think that an arithmetic simplification is at work here. What would, in that case, be the meaning of dv? We will come back to such questions when we see integration by substitution.

Derivative of the Inverse Function

We now calculate the derivative of the inverse function f^{-1} of f. We will use the notation $y = f(x)$, $x = f^{-1}(y)$. Recall that the values of x and y correspond univocally through f and f^{-1}; to each x corresponds a y, and to each y an x.

If we compose f with f^{-1} we obtain the "identity" function: $y \to y$, that is

$$y = (f \circ f^{-1})(y).$$

Differentiating (using the chain rule),

$$1 = (f \circ f^{-1})'(y) = f'(f^{-1}(y))(f^{-1})'(y).$$

From which

$$(f^{-1})'(y) = \frac{1}{f'(f^{-1}(y))} = \frac{1}{f'(x)}.$$

And we have a problem: this formula tells us the derivative of f^{-1} at the point y in terms of the derivative of f *at the point x corresponding to y*. But normally we want to express the derivative of f^{-1} in terms of its own variable, y. This is sometimes easy, and other times not so much. Let's see some examples:

Derivative of the Square Root $(\sqrt{y})' = \frac{1}{2\sqrt{y}}$.

Indeed,

$$(\sqrt{y})' = \frac{1}{(x^2)'} = \frac{1}{2x} = \frac{1}{2\sqrt{y}}.$$

Derivative of the Natural Logarithm $(\ln)'(y) = \frac{1}{y}$.
We have

$$(\ln)'(y) = \frac{1}{(e^x)'} = \frac{1}{e^x} = \frac{1}{y}.$$

Derivative of arc sine $(\arcsin)'(y) = \frac{1}{\sqrt{1-y^2}}$.

We have

$$(\arcsin)'(y) = \frac{1}{(\sin(x))'} = \frac{1}{\cos(x)}.$$

To express this in terms of the variable y: from $\cos^2 x + \sin^2 x = 1$ we have $\cos x = \sqrt{1 - \sin^2 x} = \sqrt{1 - y^2}$, and then

$$(\arcsin)'(y) = \frac{1}{\sqrt{1 - y^2}}.$$

Derivative of arc cosine $(\arccos)'(y) = \frac{-1}{\sqrt{1-y^2}}$.

$$(\arccos)'(y) = \frac{1}{(\cos(x))'} = \frac{1}{-\sin(x)}.$$

Again, from $\cos^2 x + \sin^2 x = 1$ we have $\sin x = \sqrt{1 - \cos^2 x} = \sqrt{1 - y^2}$, and then

$$(\arccos)'(y) = \frac{-1}{\sqrt{1 - y^2}}$$

Derivative of arc tangent $(\arctan)'(y) = \frac{1}{1+y^2}$,

$$(\arctan)'(y) = \frac{1}{(\tan(x))'} = \cos^2 x.$$

To write this in terms of the variable y, if in $1 = \cos^2 x + \sin^2 x$ we divide by $\cos^2 x$ we obtain,

$$\frac{1}{\cos^2 x} = 1 + \tan^2 x$$
$$= 1 + y^2.$$

Then $(\arctan)'(y) = \frac{1}{1+y^2}$.

We should note that these formulas are valid for:

$$\sqrt{\ } : (0, \infty) \to \mathbb{R}$$
$$\ln : (0, \infty) \to \mathbb{R}$$
$$\arcsin : (-1, 1) \to (-\frac{\pi}{2}, \frac{\pi}{2})$$
$$\arccos : (-1, 1) \to (0, \pi)$$
$$\arctan : \mathbb{R} \to (-\frac{\pi}{2}, \frac{\pi}{2}).$$

For example, when giving the formula for $(\arcsin)'(y)$ we solved for $\cos x$ using that it is positive (otherwise, we would have written $\cos x = -\sqrt{1 - y^2}$), but this is true because $x \in (-\frac{\pi}{2}, \frac{\pi}{2})$.

Another application of the chain rule:

The Derivative of $f(x) = x^c$ (for $x > 0$)
Applying logarithm, $\ln(x^c) = c \ln x$ and if we now differentiate:

$$\frac{1}{x^c}(x^c)' = \frac{c}{x}, \text{ from which}$$
$$(x^c)' = \frac{c}{x} x^c = c x^{c-1}.$$

The Derivative of a Parametrized Curve

We end this section with a comment on parametrized curves and their tangents. We know how to find the line tangent to the graph of a function f at the point $(a, f(a))$. It is the line that goes through that point and whose slope is $f'(a)$. But suppose that our curve is parametrized by $p : \mathbb{R} \to \mathbb{R}^2$ such that:

$$p(t) = (x(t), y(t)).$$

In other words $x(t)$ and $y(t)$ are simply the first and second coordinates of $p(t)$. How can we find the line tangent to this curve? Imagine that the curve is the graph of a function f, although this function is unknown to us. Since $p(t)$ is in the graph of f we have $y(t) = f(x(t))$. We would like to know $f'(x)$. Let's use the chain rule:

$$y'(t) = f'(x(t))x'(t).$$

From which:

$$f'(x(t)) = \frac{y'(t)}{x'(t)}.$$

In other words, the slope of the tangent to a curve parametrized by $p(t) = (x(t), y(t))$, at the point $p(t_0)$ is given by $\frac{y'(t_0)}{x'(t_0)}$ (unless of course $x'(t_0)$ is zero, but this happens only when the tangent is vertical). With the data $p(t_0)$ and $\frac{y'(t_0)}{x'(t_0)}$ we can find the tangent line, without need of knowing f. Note also that the direction of this tangent line is given by the vector $(1, \frac{y'(t_0)}{x'(t_0)})$, which is the same as the direction given by the vector $(x'(t_0), y'(t_0)) = p'(t_0)$. Thus the parametric form of the line tangent to the curve at $p(t_0)$ is $[p'(t_0)] + p(t_0)$.

Example The Circumference.

Say that we have the circumference centered at $(0, 0)$ and of radius one, and are looking for the line tangent to this circumference at the point $(\frac{1}{\sqrt{2}}, \frac{1}{\sqrt{2}})$. We know the circumference is parametrized by $p : [0, 2\pi] \to \mathbb{R}^2$ with:

$$p(t) = (x(t), y(t)) = (\cos t, \sin t)$$

and the point $(\frac{1}{\sqrt{2}}, \frac{1}{\sqrt{2}})$ is $p(\frac{\pi}{4})$. The slope of the tangent will then be:

$$\frac{y'(\frac{\pi}{4})}{x'(\frac{\pi}{4})} = \frac{\cos(\frac{\pi}{4})}{-\sin(\frac{\pi}{4})} = -1$$

and the tangent line is: $y = -x + \sqrt{2}$.

Example The Cycloid.

Recall that the cycloid is parametrized by $p[0, 2\pi] \to \mathbb{R}^2$:

$$p(t) = r(t - \sin t, -1 + \cos t).$$

We then have

$$p'(t) = r(1 - \cos t, -\sin t).$$

Its tangent at the point $p(t)$ will have slope

$$\frac{-\sin t}{1 - \cos t}.$$

We will use this later, when we consider the problem of the "brachistochrone."

First Derivative, Tangent Line, and Growth

Among the more important applications of Calculus are the problems of extrema, that is, finding maxima and minima of a function. When we talked about continuity, we proved that all continuous functions reach a maximum and a minimum on any closed interval. That theorem, however, did not tell us how to find it. It is one thing to know that something exists, and a very different thing to know where it is.

Some extrema problems may be dealt with by using the arithmetic-geometric inequality. Among the exercises there is an example. However, the notion of derivative gives us a very powerful tool applicable to any differentiable function. Intuitively, what we are interested in is clear. Imagine the graph of a function that reaches a minimum at the point c.

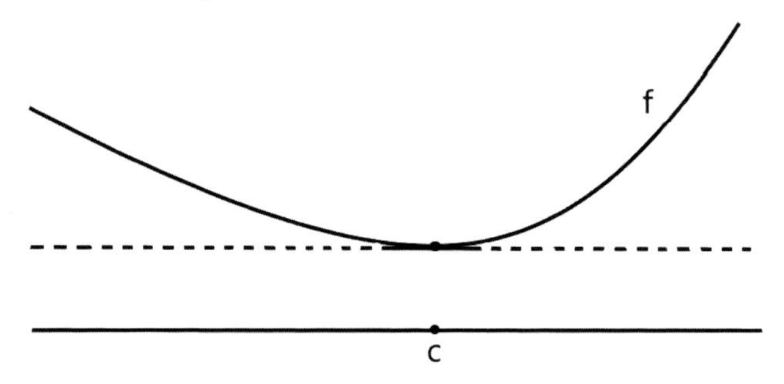

Since there the tangent line will be horizontal (the slope will be zero), we will have $f'(c) = 0$. In the proof of the following theorem, we will see that this is so. But this theorem will also allow us to prove very important related results: the Mean Value Theorems.

The Mean Value Theorems

Rolle's Theorem *If $h : [a, b] \to \mathbb{R}$ is continuous, differentiable in (a, b), and $h(a) = h(b)$, then at some point $c \in (a, b)$ the derivative of h is zero.*

Let's see. By Weierstrass' theorem h reaches a maximum and a minimum on $[a, b]$. If the maximum and the minimum coincide with $h(a) = h(b)$, we will have

$h(a) = h(x) = h(b)$ for every x; h is constant, and $h'(x) = 0$ at every point. If, however, an extreme—say a minimum—is reached at $c \in (a, b)$, let's calculate $h'(c)$. Consider the difference quotient

$$\frac{h(c+t) - h(c)}{t}.$$

The numerator is ≥ 0, for since $h(c)$ is minimum, $h(c+t) \geq h(c)$. When the denominator t is positive and tends to zero, the difference quotient will be positive, and tend to $h'(c)$, so $h'(c) \geq 0$. When the denominator t is negative and tends to zero, the difference quotient will be negative, and tend to $h'(c)$, so $h'(c) \leq 0$. Thus $h'(c) = 0$. If at c we have a maximum, the proof is analogous. □

The argument in the proof shows that if f is differentiable, and has a maximum or a minimum at c, then $f'(c) = 0$. Thus the vanishing of the derivative is a necessary (not sufficient!) condition for a differentiable function f to have an extrema at a point. The points where the derivative of f vanishes are *possible* extrema of f. Such points are called *critical points* of f.

Cauchy's Mean Value Theorem *If f and g are continuous on $[a, b]$, and differentiable in (a, b), there is a point $c \in (a, b)$ where:*

$$(f(b) - f(a))g'(c) = (g(b) - g(a))f'(c),$$

and if $g(a) \neq g(b)$,

$$\frac{f(b) - f(a)}{g(b) - g(a)} = \frac{f'(c)}{g'(c)}.$$

The proof is simple: consider the function

$$h(x) = (f(b) - f(a))g(x) - (g(b) - g(a))f(x).$$

This function is differentiable. Calculate $h(a)$ and $h(b)$:

$$\begin{aligned} h(a) &= f(b)g(a) - f(a)g(a) - g(b)f(a) + g(a)f(a) \\ &= f(b)g(a) - g(b)f(a), \end{aligned}$$

$$\begin{aligned} h(b) &= f(b)g(b) - f(a)g(b) - g(b)f(b) + g(a)f(b) \\ &= g(a)f(b) - f(a)g(b). \end{aligned}$$

Thus, $h(a) = h(b)$ and by Rolle's Theorem the derivative of h vanishes at some point c:

$$0 = h'(c) = (f(b) - f(a))g'(c) - (g(b) - g(a))f'(c),$$

thus $(f(b) - f(a))g'(c) = (g(b) - g(a))f'(c)$, and if $g(a) \neq g(b)$,

$$\frac{f(b) - f(a)}{g(b) - g(a)} = \frac{f'(c)}{g'(c)}.$$

□

Lagrange's Mean Value Theorem *If f is continuous on $[a, b]$, and differentiable in (a, b), there is a point $c \in (a, b)$ where:*

$$f(b) - f(a) = f'(c)(b - a).$$

For a proof, simply apply the previous theorem with $g(x) = x$. □

Like Rolle's Theorem, both Cauchy's and Lagrange's theorems have a clear geometric interpretation. In the case of Cauchy's consider the curve parametrized by $p : [a, b] \longrightarrow \mathbb{R}^2$, $p(t) = (g(t), f(t))$, the line through $p(a)$ and $p(b)$ has direction given by $p'(c)$

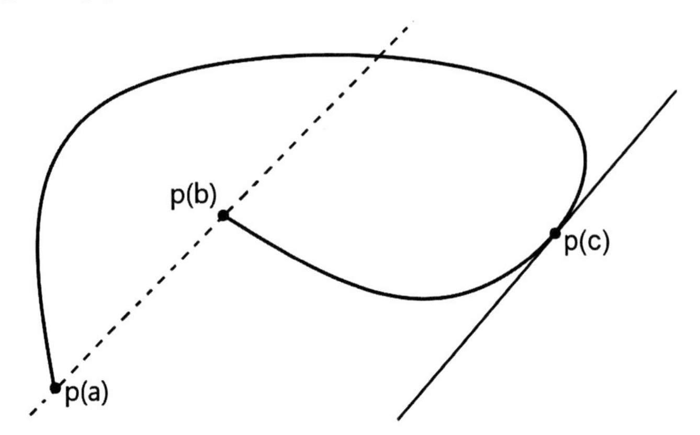

and in the case of Lagrange's, the line through $(a, f(a))$ and $(b, f(b))$ has the slope of the tangent line at $(c, f(c))$

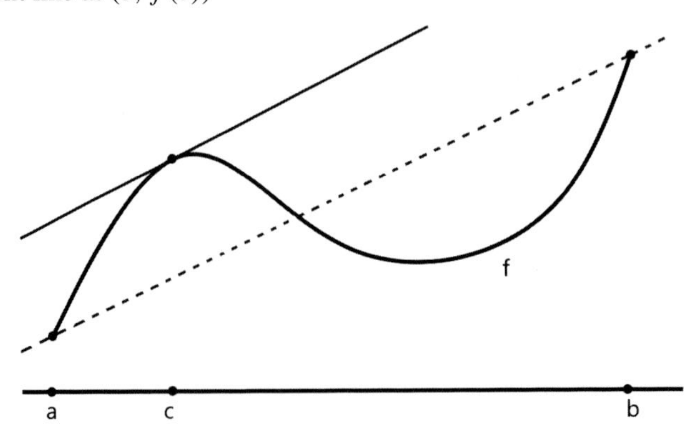

Now, if f is differentiable, and its derivative is larger than zero in the interval (a, b), it will be increasing in this interval: if $x < y$, $f(y) - f(x) = f'(c)(y - x) > 0$, for both $f'(c)$ and $y - x$ are positive. Likewise, if f has negative derivative in the interval (a, b) it will be decreasing in that interval. And if f' vanishes in (a, b), it is because the function f is constant in that interval.

Example The function $f(x) = \frac{1}{4}x^4 - x^3$ has derivative $f'(x) = x^3 - 3x^2 = x^2(x - 3)$. This derivative is zero at $x = 0$ and at $x = 3$, the critical points of f. Now, f' is < 0 for $x \in (-\infty, 0)$ and for $x \in (0, 3)$. And it is positive in $(3, \infty)$. Thus, f decreases until $x = 3$, and then grows. It will have a minimum at 3. Note that at $x = 0$ its tangent is horizontal, but there is no extrema there:

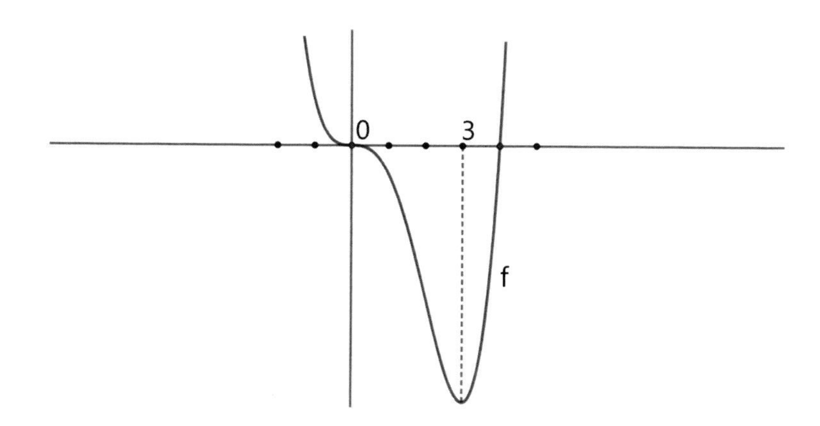

L'Hôpital's Rule

Positive derivative indicates growth; a negative one, a decline; and the size of this derivative (which is the slope of the tangent line) a larger or lesser velocity in that growth. Sometimes, to compare the velocity of growth of a function f with that of g one uses the quotient $\frac{f}{g}$. For example, which function tends more rapidly to zero (as $x \to 0$), $f(x) = x^2$ or $g(x) = x$? Setting $\frac{f(x)}{g(x)} = \frac{x^2}{x} = x \to 0$, we see that the numerator decreases faster. But what if it is not possible to "simplify" as we have done in the second equality? Suppose for a moment that $f(a) = g(a) = 0$, and that both f and g are differentiable at a. Then

$$\frac{f(x)}{g(x)} = \frac{f(x) - f(a)}{g(x) - g(a)} = \frac{\dfrac{f(x) - f(a)}{x - a}}{\dfrac{g(x) - g(a)}{x - a}},$$

and when $x \to a$,

$$\lim_{x\to a} \frac{f(x)}{g(x)} = \frac{f'(a)}{g'(a)}.$$

This is known as L'Hôpital's Rule. However, our hypotheses are excessive. With more work we can obtain the following stronger version which does not even require that the functions be defined at a.

L'Hôpital's Rule (Bernoulli) *If f and g are defined and differentiable near a, and*

$$\lim_{x\to a} f(x) = \lim_{x\to a} g(x) = 0 \quad \text{(A)}$$

$$\text{or } \lim_{x\to a} f(x) = \lim_{x\to a} g(x) = \infty \quad \text{(B)},$$

and $\lim_{x\to a} \frac{f'(x)}{g'(x)} = L$ *exists, then* $\lim_{x\to a} \frac{f(x)}{g(x)} = L$. Let's see why: consider the interval (x, a) and set

$$m(x) = \inf \frac{f'(c)}{g'(c)} \quad \text{for } c \in (x, a), \text{ and}$$

$$M(x) = \sup \frac{f'(c)}{g'(c)} \quad \text{for } c \in (x, a).$$

For each y with $x < y < a$, by the Mean Value Theorem (Cauchy), there is a c such that $x < c < y$ with (in case (A)):

$$m(x) \le \frac{f'(c)}{g'(c)} = \frac{f(x) - f(y)}{g(x) - g(y)} = \frac{\dfrac{f(x)}{g(x)} - \dfrac{f(y)}{g(x)}}{1 - \dfrac{g(y)}{g(x)}} \le M(x).$$

$$\text{Having } y \to a, m(x) \le \frac{f(x)}{g(x)} \le M(x), \text{ by (A).}$$

$$\text{And now having } x \to a, L \le \lim_{x\to a} \frac{f(x)}{g(x)} \le L.$$

In case (B),

$$m(x) \le \frac{f'(c)}{g'(c)} = \frac{f(y) - f(x)}{g(y) - g(x)} = \frac{\dfrac{f(y)}{g(y)} - \dfrac{f(x)}{g(y)}}{1 - \dfrac{g(x)}{g(y)}} \le M(x).$$

$$\text{Having } y \to a, m(x) \le \liminf_{y\to a} \frac{f(y)}{g(y)} \le M(x),$$

$$\text{and now when } x \to a,\ L \le \liminf_{y\to a} \frac{f(y)}{g(y)} \le L.$$

And analogously for lim sup (see Exercise 16 of Chap. 3). In both cases, $\lim_{x\to a} \frac{f(x)}{g(x)} = L = \lim_{x\to a} \frac{f'(x)}{g'(x)}$. □

Example $\lim_{x\to 0} x \ln x$:

$$x \ln x = \frac{\ln x}{\frac{1}{x}} \to \frac{\infty}{\infty}$$

$$\frac{f'}{g'}: \quad \frac{\frac{1}{x}}{-\frac{1}{x^2}} = -\frac{x^2}{x} = -x \to 0,$$

then $\lim_{x\to 0} x \ln x = 0$.

Among the exercises we'll see more examples. We should comment here on a notation commonly used in relation to "speed of decay" or "speed of decline." We will write $f(x) = \circ(g(x))$, if $\frac{f(x)}{g(x)}$ tends to zero as x tends to zero. This helps to express the speed of decay of f: we write $f(x) = \circ(x)$ if f decreases faster than x, and $f(x) = \circ(x^2)$ if f decreases more rapidly than x^2.

Snell's Law

Let's start with a typical extrema problem: we want to go, in the least possible time, from point A on one bank of a river to point B on the opposite bank. The river is a kilometers wide, and point B is c kilometers further East.

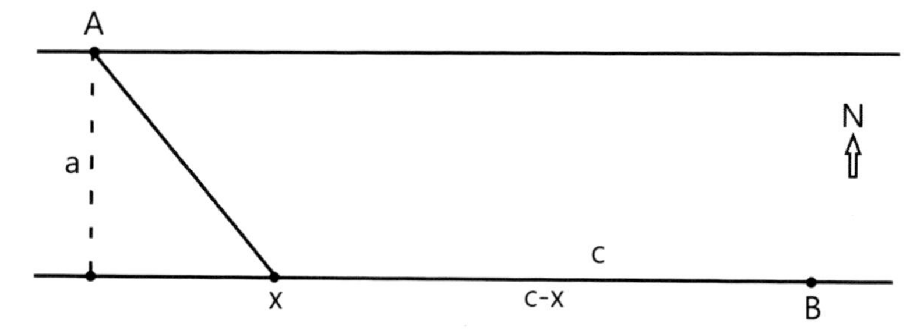

We have a rowboat to cross the river, and we can row at 5 km/h, but—once on the other side—we can run at 13 km/h. Where do we disembark? We can disembark at any point, say x km East of A, and run the rest of the way ($c - x$ kilometers). How do we choose x to minimize the time?

Let's express the time T as a function of x. Recall that speed $= \frac{\text{space}}{\text{time}}$, so time $= \frac{\text{space}}{\text{speed}}$.

If we disembark at point X we row $\sqrt{x^2 + a^2}$ km at 5 km/h, and then run $c - x$ km at 13 km/h. This will take us:

$$\begin{aligned} T(x) &= \text{time rowing + time running} \\ &= \frac{\sqrt{x^2 + a^2}}{5} + \frac{c - x}{13}. \end{aligned}$$

To find the minimum of this function, we look for the value of x for which the derivative vanishes.

$$T'(x) = \frac{x}{5\sqrt{x^2 + a^2}} - \frac{1}{13} = 0$$

that is,

$$\begin{aligned} 13x &= 5\sqrt{x^2 + a^2} \\ 169x^2 &= 25(x^2 + a^2) \\ (169 - 25)x^2 &= 25a^2 \\ 144x^2 &= 25a^2 \\ 12x &= 5a \\ x &= \frac{5a}{12}. \end{aligned}$$

One may verify that $T' < 0$ in $(0, \frac{5a}{12})$ and $T' > 0$ in $(\frac{5a}{12}, \infty)$, so at $\frac{5a}{12}$ we do have a minimum. We shall disembark at $x = \frac{5a}{12}$.

We now change our problem by setting point B, b km South of its original location:

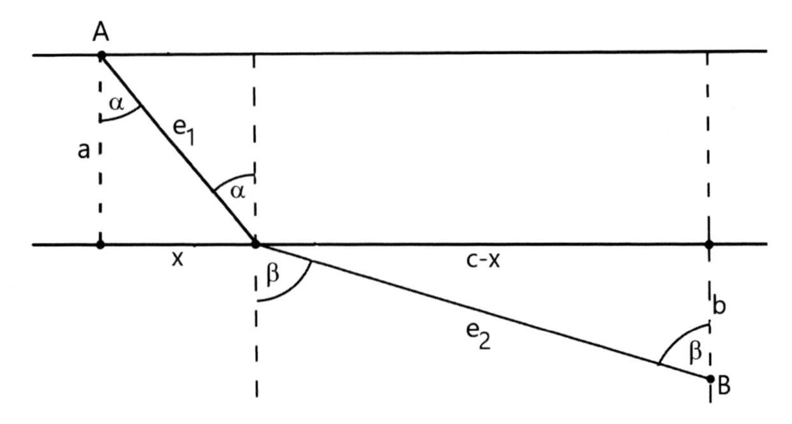

The problem is more complicated. But now our intention will not be to find x, but rather to make a geometric comment. Our function $T(x)$ is now (say our speed in the water is v_1 and on land, v_2):

$$T(x) = \frac{e_1}{v_1} + \frac{e_2}{v_2} = \frac{\sqrt{x^2 + a^2}}{v_1} + \frac{\sqrt{(c-x)^2 + b^2}}{v_2}.$$

Differentiating,

$$\begin{aligned} T'(x) &= \frac{x}{v_1\sqrt{x^2 + a^2}} - \frac{c-x}{v_2\sqrt{(c-x)^2 + b^2}} \\ &= \frac{\frac{x}{e_1}}{v_1} - \frac{\frac{c-x}{e_2}}{v_2} \\ &= \frac{\sin\alpha}{v_1} - \frac{\sin\beta}{v_2}. \end{aligned}$$

The time will be minimized when $\frac{\sin\alpha}{v_1} = \frac{\sin\beta}{v_2}$. This is called *Snell's Law*. It is the law that light rays must obey while passing through media of different density, in which their speed is different (larger density, lesser speed). This is the cause of the "refraction" of light and explains why submerged objects seem to us to be where they are not.

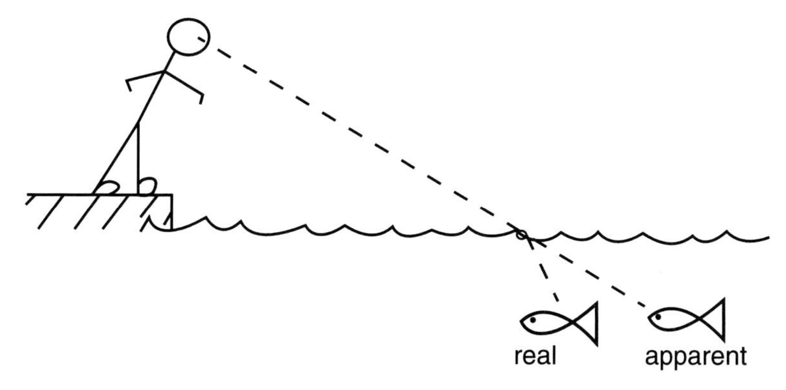

The name *Snell's Law* is linked to this property of refraction of light. However, it is useful to remember that it is not a physical law, but a mathematical one which must be obeyed by any object if it is to go from one point to another in the least possible time. We will use it now to solve the "Brachistochrone problem." This law has the name of the XVIth Century Dutch mathematician, Willebrord Snell. As often happens in the history of Calculus, it was already known to the Xth Century Persian mathematician Ibn Sahl.

The Brachistochrone

Johann Bernoulli (1667–1748) was, as was his brother Jacob, one of the first to study the consequences of Leibniz' calculus. The two brothers competed, often bitterly, to obtain results. The brachistochrone problem, posed by Johann, was one of the first to pave the way to what today is the calculus of variations.

Johann set, in 1696, the following challenge to "the most brilliant mathematicians in the world":

> "Given two points A and B in a vertical plane, what is the curve traced out by a point acted on only by gravity, which starts at A and reaches B in the shortest time?"

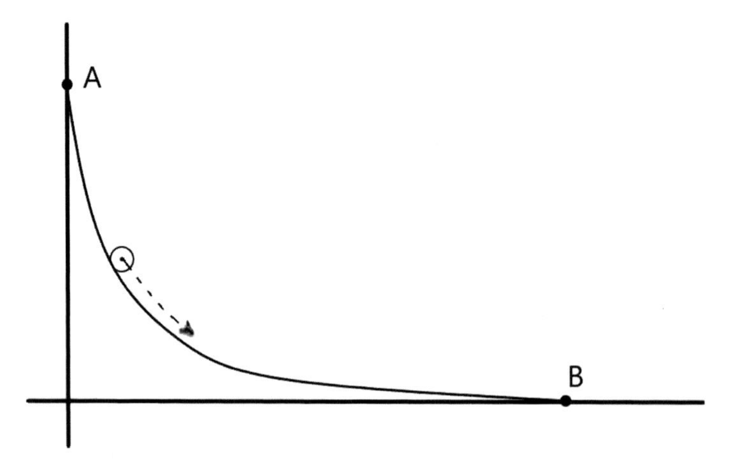

Such a curve is called "brachistochrone" (brachistos = "shortest"; cronos = "time"). Today there are many people in the world that can solve this problem. In 1696 there were six: Johann Bernoulli himself, his brother Jacob, L'Hôpital, Leibniz, Newton, and von Tschirnhaus (Galileo had tried it in 1638, unsuccessfully).

Let's find which curve it is. Say it is parametrized by $p(t) = (x(t), y(t))$, (that is, y is the height of the point). Recall from Physics courses that:

$$\text{Potential Energy} = mgy(t)$$

$$\text{Kinetic Energy} = \frac{1}{2}mv(t)^2,$$

where m is the point's mass, g is gravity's acceleration (which we consider positive), and v the point's velocity. As the point falls, its potential energy is transformed into kinetic energy, but the principle of conservation of energy tells us that:

$$\text{Potential Energy} + \text{Kinetic Energy} = \text{constant} = mgy(0)$$

$$mgy(t) + \frac{1}{2}mv(t)^2 = mgy(0)$$

$$\frac{1}{2}mv(t)^2 = -mg(y(t) - y(0))$$
$$v(t)^2 = -2g(y(t) - y(0)).$$

At any given moment our point is moving along the curve $(x(t), y(t))$, whose slope is $\frac{y'(t)}{x'(t)} = -\frac{a}{b}$.

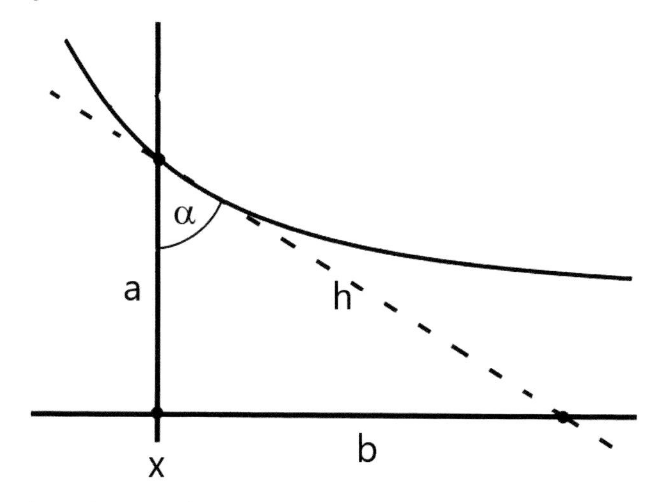

And as the point goes from one point to another in the least possible time, it obeys Snell's Law: its velocity v is such that:

$$\begin{aligned} c &= \frac{v}{\sin\alpha} \qquad \text{is constant} \\ &= \frac{vh}{b}. \\ \text{Then} \quad c^2 &= \frac{v^2h^2}{b^2} \\ &= v^2\frac{b^2+a^2}{b^2} \\ &= v^2\left(1+\left(\frac{a}{b}\right)^2\right) \\ &= v^2\left(1+\frac{y'(t)^2}{x'(t)^2}\right). \end{aligned}$$

In other words

$$(y(t) - y(0))\left(1+\frac{y'(t)^2}{x'(t)^2}\right) = k$$

must be a constant.

This is a *differential equation*: an equation involving the derivatives of unknown functions x and y. We want to find the functions that verify this equation (just as $f(x) = e^x$ verifies the equation $f - f' = 0$). As it turns out, the cycloid verifies the brachistochrone equation. Recall that we parametrized it as follows:

$$p(t) = r(t - \sin t, \cos t - 1),$$

from where

$$p'(t) = r(1 - \cos t, -\sin t), \text{ and then: } \frac{y'(t)}{x'(t)} = \frac{-\sin t}{1 - \cos t}.$$

But

$$\begin{aligned}
(y(t) - y(0))\left(1 + \frac{y'(t)^2}{x'(t)^2}\right) &= r(\cos t - 1)\left(1 + \frac{\sin^2 t}{(1 - \cos t)^2}\right) \\
&= r(\cos t - 1)\left[\frac{(1 - \cos t)^2 + \sin^2 t}{(1 - \cos t)^2}\right] \\
&= r(\cos t - 1)\left[\frac{1 - 2\cos t + \cos^2 t + \sin^2 t}{(1 - \cos t)^2}\right] \\
&= r(\cos t - 1)\left[\frac{2(1 - \cos t)}{(1 - \cos t)^2}\right] \\
&= \frac{2r(\cos t - 1)}{(1 - \cos t)} \\
&= -2r, \qquad \text{which is constant.}
\end{aligned}$$

Christian Huygens had posed (and solved) in 1659 the *tautochrone* problem (tauto = same, cronos = time): to find the curve that a point would move along (to its end) in the same amount of time, no matter where it starts. The tautochrone is (again!) the cycloid.

Exercises

1 Write the equation of the line that:

(i) Contains the point $(a, f(a))$, and
(ii) has slope $f'(a)$.

2 Use the Newton–Raphson method to find (approximately):

(i) Zeros of $f(x) = x^3 - 3x + 4$,
(ii) Zeros of $f(x) = x^4 + x^2 - 6x + 2$,
(iii) $\sqrt{10}$.

3 If L is the line tangent to the graph of f at $(x, f(x))$, What is the point of intersection of L and the x-axis?

4 Using the difference quotient, calculate the derivative at point a of:

(i) the constant function $f(x) = c$,
(ii) the function $f(x) = x^2$,
(iii) the function $f(x) = \sqrt{x}$ (Hint: $1 = \frac{\sqrt{a+h}+\sqrt{a}}{\sqrt{a+h}+\sqrt{a}}$).

5 The function f has the property: $|f(x)| \le x^2$ for all x. Prove that f is differentiable at $a = 0$, and $f'(0) = 0$.

6 Let $f(x) = |x|$. Prove that f is differentiable at a if $a > 0$ or if $a < 0$, but not if $a = 0$.

7 Suppose f and g are differentiable. Prove the formulas for

(i) Derivative of a sum: $(f + g)'(x) = f'(x) + g'(x)$.
(ii) Derivative of a product: $(fg)'(x) = f'(x)g(x) + f(x)g'(x)$.
(Hint: $0 = -f(x)g(x + h) + f(x)g(x + h)$).
(iii) Derivative of a quotient: $\left(\frac{f}{g}\right)'(x) = \frac{f'(x)\cdot g(x) - f(x)g'(x)}{g(x)^2}$.
(Hint: $0 = -f(x)g(x) + f(x)g(x)$).

8 Prove the following formulas for the derivatives of the hyperbolic functions.

$$\begin{aligned}(\sinh x)' &= \cosh x\\ (\cosh x)' &= \sinh x\\ (\tanh x)' &= \frac{1}{(\cosh x)^2}.\end{aligned}$$

9 Given the function $f : (r_1, r_2) \longrightarrow \mathbb{R}$,

$$f(x) = -x^2 + (r_1 + r_2)x - r_1 r_2 = (r_2 - x)(x - r_1),$$

find its maximum,

(i) without using derivatives (use the arithmetic-geometric inequality: $\sqrt{ab} \le \frac{a+b}{2}$).

(ii) using derivatives.

10 Find two positive numbers whose sum is 9 and have the largest possible product.

11 Prove that for all $x \geq 1$, $2\sqrt{x} - 2 \leq x - 1$.

12 Of all the points in the graph of $y = \sqrt{2x}$, which is closest to $(4, 0)$? At what distance is it?

13 The ellipse of radii a and b is given by the equation

$$\frac{x^2}{a^2} + \frac{y^2}{b^2} = 1.$$

What is the rectangle (with sides parallel to the axes) of largest area which can be inscribed in this ellipse?

14 We wish to construct a rectangular window, topped by a semicircle. If the perimeter is to be 4 meters, what will its dimensions be if the rectangular part is to have the largest possible area?

15 We have a rectangle with a 12cm perimeter. A cylinder is obtained by rotating it around one of its sides. What dimensions must the rectangle have in order for the cylinder to have the largest possible volume?

16 What dimensions (radius and height) will a cone have if it is the cone of largest volume inscribed in a sphere of radius 2?

17 What dimensions (radius and height) will a cylinder have if it is the cylinder of largest volume inscribed in a sphere of radius 2?

18 Of all the segments that go from the x-axis to the y-axis and contain the point $P = (a, b)$ (with $a > 0$ and $b > 0$), which is the shortest?

19 A rectangular sheet of paper is 20cm high and 10cm wide. We fold it by taking the bottom right corner to the left side of the sheet. What is the least possible length the fold can have?

20 Reflection of light. Observe the following diagram:

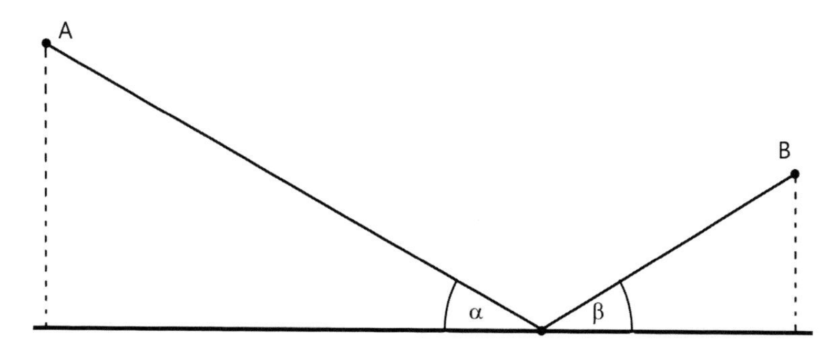

A ray of light must go from point A to a mirror, and then to point B. What will be the relation between angles α and β? (NOTE: light wants to go as fast as possible).

21 Bearing in mind that

$$\ln(a^x) = x \ln a,$$

(a) Differentiate both sides of the equality.
(b) Solve for $(a^x)'$.

What is the derivative of $f(x) = a^x$?

22 A cylindrical water tank has a radius of 40cm and a height of one meter. It is being filled at a rate of 4 litres per minute. What is the rate of change of the height of the water in the tank?

23 Another water tank is shaped like an inverted cone, 2 meters high and with a radius of one meter (at its roof). It is being filled at a rate of 4 litres per minute.

(a) What is the rate of change of the height of the water in the tank when this height is 50cm?
(b) What is the rate of change of the height of the water in the tank when this height is 1,20m?

24 Given the function

$$f(x) = \begin{cases} x^2 \sin \frac{1}{x}, & \text{if } x \neq 0 \\ 0, & \text{if } x = 0, \end{cases}$$

(i) prove that $f'(0) = 0$ (Hint: Exercise 5),
(ii) calculate $f'(x)$ for $x \neq 0$,

(iii) Is f' continuous at $x = 0$?

25 *Order of decay*: We will say that $f(x) \to 0$ "as $(x-a)^n$" when $x \to a$, if

$$\lim_{x \to a} \frac{f(x)}{(x-a)^n} = c, \text{ a } \mathbf{non-zero} \text{ real number.}$$

We will say that $f(x)$ is as "o $(x-a)^n$" (and write $f(x) = \circ(x-a)^n$), if

$$\lim_{x \to a} \frac{f(x)}{(x-a)^n} = 0.$$

Study the order of decay of the following functions at the points indicated in each case.

(i) $f(x) = \sin^3 x$ at the point $a = 0$.
(ii) $f(x) = (\ln x)^2$ at the point $a = 1$.
(iii) $f(x) = (x-2)^3 \cos x$ at the point $a = 2$.
(iv) $f(x) = (\cos x - 1)^2$ at the point $a = 0$.

26 Show that the difference between f and its tangent line at a:

$$f(x) - \left[f(a) + f'(a)(x-a)\right]$$

is as $\circ(x-a)$.

The Integral 5

We now consider the notion of integral, and its relation to areas. The Fundamental Theorem of Calculus places the integral as an antiderivative and allows for some consequences of our knowledge of derivatives to integration. We give a short presentation of uniform convergence. The length of a parametrized curve is also defined.

Measure and Integral

The idea of "integration" is as old as mathematics: it starts with the necessity of measuring areas delimited by curves. The problem is that our units of surface area, for example, square meters or square feet, do not fit well on such surfaces and this makes comparison difficult. However, Greek mathematicians invented methods to calculate exactly areas and volumes of a wide range of geometric figures. As we have seen in Chap. 3, Antiphon, Eudoxus, and Archimedes used successfully the idea of "exhaustion" to fill up an area they wished to measure with objects whose measure they knew, and then used some notion of limit to obtain the total area. Archimedes, and in the XVIIth Century, Cavalieri, used the idea of area as a *sum of lines* which we will see in Chap. 8. The contemporaries of Newton and Leibniz did not doubt the existence of an "area under a curve" and used the Fundamental Theorem of Calculus to calculate it. In a few pages we will go down that path. But before that, I would like to stop and tell you about some of the notions of integral and to show you some of the difficulties implicit in the subject (which we will avoid!).

In the mid-nineteenth century Calculus was needing a formalization of many of its concepts. Among them the integral, which was being used by mathematicians since the XVIIth Century. There have appeared since then various definitions of integral: the Riemann integral, the Darboux integral, the Riemann–Stieltjes integral, the Lebesgue integral, the Daniell integral, the Henstock–Kurzweil integral... To understand some of these ideas and to comprehend the origin and importance of the

I. Zalduendo, *Calculus off the Beaten Path*, SUMS Readings,
https://doi.org/10.1007/978-3-031-15765-3_5

notion of "measure" we will comment only two of them: the Riemann integral and the Lebesgue integral.

The Riemann Integral (1854)
Consider a positive function f defined on some closed interval $[a, b]$. We wish to find the area under the graph of f. We split the interval $[a, b]$ into n sub-intervals using $a = x_0 < x_1 < \cdots < x_n = b$ and we take in each small interval a point $t_i \in [x_i, x_{i+1}]$. Then we define the *Riemann sum* corresponding to this selection of x_i's and t_i's:

$$\sum_{i=0}^{n-1} f(t_i)(x_{i+1} - x_i).$$

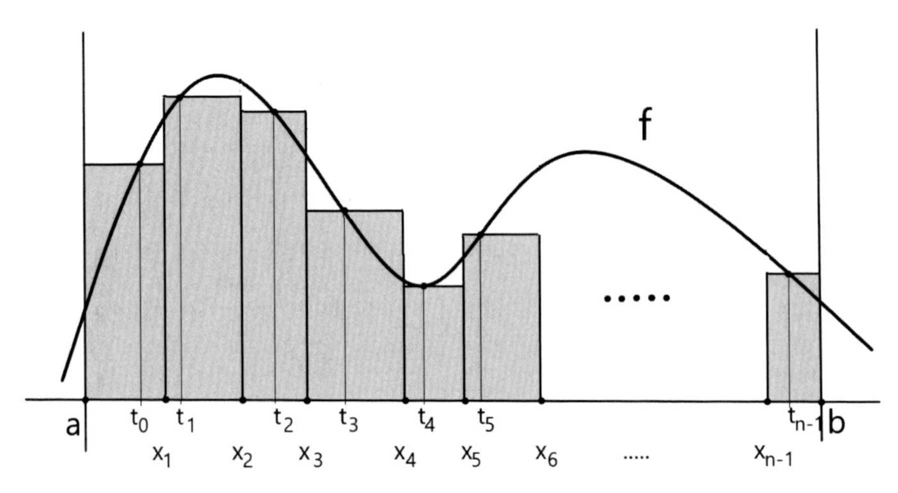

As we see in the picture, the Riemann sum corresponds to the sum of areas of rectangles with base $[x_i, x_{i+1}]$ and height $f(t_i)$. One would expect that if we make the intervals on the bases smaller and smaller, the area will adjust to the area under the graph of f. The function f will be said to be *Riemann integrable* and its integral will be the number I when this happens: for each $\varepsilon > 0$ there exists a $\delta > 0$ which assures that if $(x_{i+1} - x_i) < \delta$ for every i, then

$$\left|\sum_{i=0}^{n-1} f(t_i)(x_{i+1} - x_i) - I\right| < \varepsilon.$$

This happens whenever f is continuous, and also sometimes when it is not. However, it is easy to find functions that are not Riemann integrable, for example, the indicator function of the rational numbers:

$$f(x) = \begin{cases} 1, & \text{if } x \in \mathbb{Q} \\ 0, & \text{if } x \notin \mathbb{Q}. \end{cases}$$

The Lebesgue Integral (1904)

Consider the same function as above, but now instead of partitioning the interval $[a, b]$, we partition the range of the function with $a_1 < a_2 < a_3 < \cdots < a_n$, and consider the subsets of $[a, b]$ given by

$$A_k = \{x \in [a, b] : a_k \leq f(x) < a_{k+1}\},$$

as in the picture

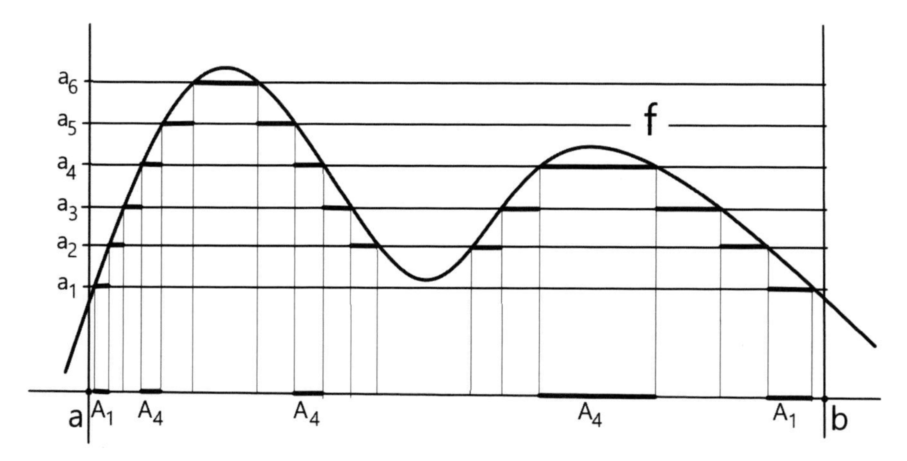

Now consider the sums

$$\sum_{k=1}^{n-1} a_k m(A_k),$$

where $m(A_k)$ is the measure of the set A_k. The sets A_k are in general not intervals, as can be seen in the picture. One expects that these sums will converge to the integral of f when you partition the range of f more finely (the definition is actually a bit more complicated). In any case, the Lebesgue integral requires the development a *measure theory* in order to consider the measure $m(A_k)$ for a wide range of possible subsets $A_k \subset [a, b]$. This is not a disadvantage but a strength of the Lebesgue integral, for by enlarging the class of subsets whose measures can be used we also enlarge the class of functions which may be integrated. For example, in the case of the indicator function of the rational numbers, for $a_1 = \frac{1}{2}$ and $a_2 = 1$, one has $A_1 = \mathbb{Q} \cap [a, b]$ which measures 0, and in fact the function is

Lebesgue integrable and its integral is 0. Every Riemann integrable function is also Lebesgue integrable. But the true strength of the Lebesgue integral is not that there are more integrable functions, but rather the existence of very strong and easy to use theorems which permit the interchange of the integral with limits of a sequence of functions ($\int \lim f_n = \lim \int f_n$). Also, the Lebesgue integral may be defined in many situations—where there is an adequate theory of measure—over sets which may be more abstract than $[a, b]$, for example, probability spaces. We will not enter here into these subjects. To end this discussion, I want to show one of the difficulties in the idea of more general integrals.

A Non-measurable Set

One of the requirements of measure theory is that if m is a measure, and $A_1, A_2, A_3, \ldots$ are countably many disjoint sets (i.e., $A_i \cap A_j = \emptyset$ for $i \neq j$), which can be measured, then

$$m\left(\bigcup_k A_k\right) = \sum_k m(A_k),$$

in other words the measure of the union is the sum of the measures. We will see now that for the usual measure on the real line (for which $m([a, b]) = b - a$) there are subsets which are non-measurable. We will construct one such set: the Vitali set.

Define for real numbers x, y, the relation: $x \sim y$ if $x - y$ is rational. This is an *equivalence relation*, in other words:

(i) $x \sim x$: indeed, $x - x = 0$, which is rational.
(ii) If $x \sim y$, then $y \sim x$: if $x - y$ is rational, so is $y - x$.
(iii) If $x \sim y$ and $y \sim z$, then $x \sim z$: if $x - y$ and $y - z$ are rational, so is $x - z = (x - y) + (y - z)$.

The *equivalence class* of a number x is $\{y : x \sim y\}$ that is, the set of all elements which are equivalent to x. We construct the Vitali set $V \subset [0, 1]$ by taking exactly *one* element from each equivalence class. We will suppose—towards a contradiction—that V is measurable, and call $m(V)$ its measure. Now if $\mathbb{Q} \cap [-1, 1] = \{r_1, r_2, r_3, \ldots\}$ (remember that $\mathbb{Q} \cap [-1, 1]$ is countable), we set

$$V_k = V + r_k.$$

These are translates of V, so they measure the same: $m(V_k) = m(V)$. Note also that $V_i \cap V_j = \emptyset$ for $i \neq j$: if $x \in V_i \cap V_j$, we would have $v + r_i = x = v' + r_j$, and then $v - v' = r_j - r_i$, a rational number. But then we would have $v \sim v'$, but as in V there's only one element of each equivalence class, this means that $v = v'$ and therefore $r_i = r_j$, which is absurd, for $i \neq j$. Let's now see that

$$[0,1] \subset \bigcup_k V_k \subset [-1,2].$$

The first inclusion is because if $x \in [0,1]$, there exists $v \in V$ such that $x \sim v$, in other words $x - v \in \mathbb{Q} \cap [-1,1]$; $x - v = r_k$ for some k, and then $x = v + r_k \in V_k$. The second inclusion is because for each k, $V_k = V + r_k \subset [0,1] + [-1,1] = [-1,2]$. Then

$$m([0,1]) \leq m\left(\bigcup_k V_k\right) = \sum_k m(V_k) \leq m([-1,2]),$$

that is,

$$1 \leq \sum_k m(V) \leq 3,$$

which cannot happen if $m(V) = 0$, or if $m(V) > 0$. Thus $m(V)$ does not exist: V is not measurable.

The Area Under a Curve

Let's return now to the innocence of the XVIIth Century, and consider a continuous function $f : [a,b] \to \mathbb{R}$ taking positive values. I will ask you to accept that the area of the region under the graph of f really exists. It does, because all continuous functions on $[a,b]$ are Riemann integrable. The area under the graph could be calculated using the Riemann integral (but won't be):

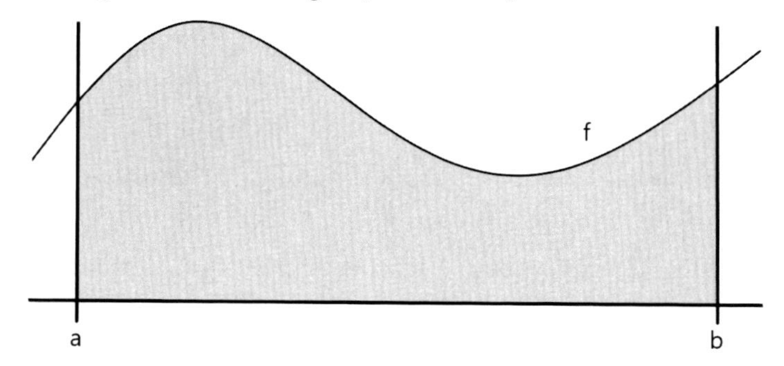

This area is a number which we will call *the integral* of f between a and b and which we will denote by:

$$\int_a^b f \qquad \text{or by} \qquad \int_a^b f(x)\,dx.$$

It is clear that if $a \leq c \leq b$, $\int_a^b f = \int_a^c f + \int_c^b f$,

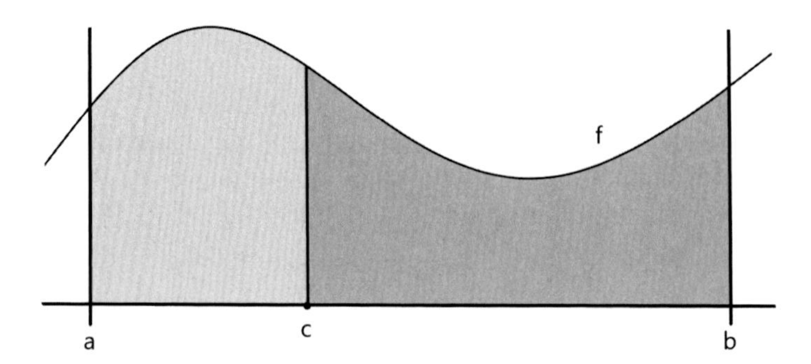

in other words: the area between a and b is equal to the area between a and c plus the area between c and b. This, for functions $f \geq 0$. If f is negative in some interval $[a, b]$, we define its integral as the negative number:

$$\int_a^b f = -\int_a^b (-f).$$

Note that $(-f) \geq 0$. We may consider that with this definition, the areas under the x axis are subtracted. For example, $\int_0^{\pi} \cos x \, dx = 0$:

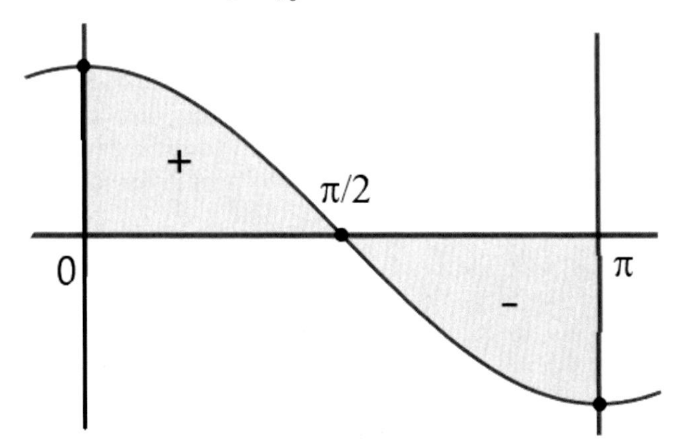

The interesting thing would be to be able to calculate, for example, the area $\int_0^{\frac{\pi}{2}} \cos x \, dx$. We have mentioned that the Greeks were able to calculate many areas, in general approximating them by simpler regions whose areas they knew. But in the XVIIth Century something truly extraordinary happened: the idea of integral met the idea of derivative, and this gave rise to the Fundamental Theorem of Calculus.

The Fundamental Theorem of Calculus

Admitting, as by now should be clear, that Calculus is a set of ideas and concepts that crystallized along many centuries, Newton and Leibniz are usually assigned the "invention" of Calculus. In both mathematicians (and many books have been written about the priority dispute) appear the general ideas of derivative and integral, and more importantly, the connection between them: the Fundamental Theorem of Calculus, although Barrow who was Newton's teacher, and James Gregory had already seen it in particular cases.

Gottfried Leibniz (1646–1716), was born in Leipzig, Germany. He was a mathematician and philosopher, but he also made important contributions to other disciplines, among them physics, biology, and history. The diversity and extension of his work is enormous. He was also the inventor of several mechanical calculating machines.

Isaac Newton (1643–1727), was born in Woolsthorpe, England. In 1665 Cambridge University, where Newton studied, was closed due to an epidemic of bubonic plague. Newton went back home to Woolsthorpe, where in less than two years he invented calculus and made other important discoveries in optics, physics, and astronomy. He was not yet 25. Twenty years later, in 1687, he published these and other results in his great masterpiece Mathematical principles of natural philosophy, *commonly called the* Principia. *There he describes the universal law of gravitation, sets the foundations of classical mechanics, and applies his results to the movement of the planets and the tides.*

Isaac Barrow (1630–1677), was a Professor at Cambridge. He occupied his chair only six years, and was succeeded by Newton in 1669. In his Lectures on Geometry *he gives a version of the Fundamental Theorem of Calculus.*

Say that $f : \mathbb{R} \to \mathbb{R}$ is continuous, and fix $a \in \mathbb{R}$. Then, for any $x > a$ we may consider $\int_a^x f$. This defines a new function $F : [a, \infty) \to \mathbb{R}$ setting:

$$F(x) = \int_a^x f.$$

That is, for each x, $F(x)$ is the area under the graph of f between a and x.

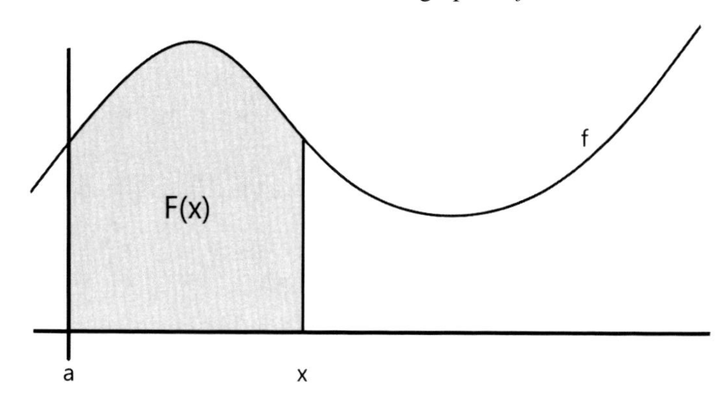

If one asks how F increases or decreases when we move x, it is reasonable to think that if f is large, F will grow rapidly; if f is small, F will grow more slowly; if f is negative F will decrease (we will be subtracting areas). But more precisely:

Fundamental Theorem of Calculus *If f is continuous and $F(x) = \int_a^x f$, then F is differentiable and $F' = f$.*

Let's prove the theorem: we must consider the quotient $\frac{F(x+h)-F(x)}{h}$, and see what happens when h tends to zero. Look at the following graph and consider

$$F(x+h) - F(x) = \int_x^{x+h} f.$$

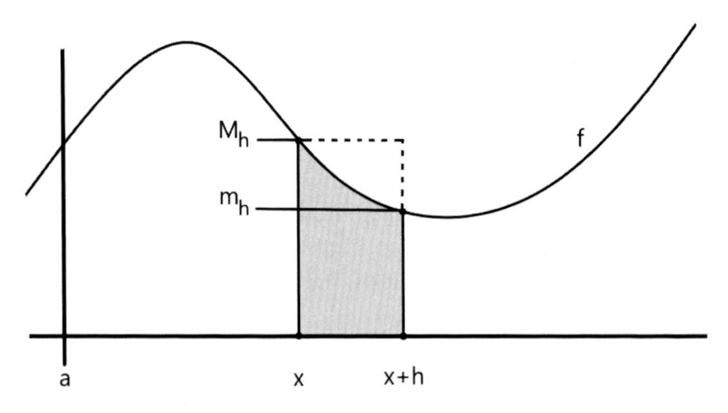

That is, the area under f between x and $x+h$. If the minimum of f in $[x, x+h]$ is m_h, and the maximum of f in $[x, x+h]$ is M_h, clearly the area under f will be between the areas of the two rectangles:

$$m_h h \leq F(x+h) - F(x) \leq M_h h$$

So

$$m_h \leq \frac{F(x+h) - F(x)}{h} \leq M_h.$$

What will happen when h tends to zero? The interval $[x, x+h]$ becomes smaller and smaller and, since f is continuous, all values taken by f in this interval (including m_h and M_h) become closer and closer to $f(x)$:

$$f(x) = \lim_{h \to 0} m_h \leq \lim_{h \to 0} \frac{F(x+h) - F(x)}{h} \leq \lim_{h \to 0} M_h = f(x).$$

Thus $F'(x) = f(x)$, and we are done. □

Let's say now that we want to calculate $\int_a^b f$. This is $F(b)$, but we do not know F. But suppose that we know some function G whose derivative is f; that is $G' = f$. This is what we call a *primitive* of f. Since F is also a primitive of f, for $F' = f$, it turns out the G and F will differ by a constant. Thus to know G will be almost like knowing F. This is the content of the following Corollary to the Fundamental Theorem of Calculus.

Barrow's Rule *If G is a primitive of f,*

$$\int_a^b f = G(b) - G(a).$$

Let's see why. Since both G and F are primitives of f,

$$(G - F)' = G' - F' = f - f = 0.$$

Then, $G - F$ is a constant, k. So, $G = F + k$. Now,

$$G(b) - G(a) = (F(b) + k) - (F(a) + k) = F(b) - F(a) = F(b) = \int_a^b f.$$

□

Example We wish to calculate $\int_0^{\frac{\pi}{2}} \cos x \, dx$. And we know that $(\sin x)' = \cos x$. Then by Barrow's rule:

$$\int_0^{\frac{\pi}{2}} \cos x \, dx = \sin\left(\frac{\pi}{2}\right) - \sin(0) = 1.$$

Great! And we have a long list of primitives... as long as our list of derivatives. The notation $G(x)\Big|_a^b$ is usually used to indicate $G(b) - G(a)$. Thus, for example,

$$\int_1^b \frac{1}{x} \, dx = \ln x \Big|_1^b = \ln b - \ln 1 = \ln b$$

$$\int_0^b x^2 \, dx = \frac{1}{3}x^3 \Big|_0^b = \frac{1}{3}b^3 - \frac{1}{3}0^3 = \frac{1}{3}b^3.$$

A Pause for Comments

First, a comment on notation. We have defined the integral only for functions f continuous on a closed interval $[a, b]$. The "ingredients" of our definition are the function f and the interval $[a, b]$. Thus, economy of design would indicate that these should also be the ingredients of our notation:

$$\int_a^b f.$$

This is fine for abstract functions such as f. However, many of our more concrete functions simply cannot be named without reference to a variable: we do not say "the function f which assigns to each real number its square," we say x^2. Similarly, we do not write "cos," we write $\cos t$. When integrating such functions we will use the standard notation, as in

$$\int_0^1 x^2\,dx \qquad \text{and} \qquad \int_0^{\pi/2} \cos t\,dt.$$

The notation $\int_a^b f(x)\,dx$ also helps to indicate which is the variable of integration (the variable of the function we are integrating) in situations where there could be some ambiguity, such as

$$\int_0^2 xt^2\,dx = 2t^2 \qquad \text{and} \qquad \int_0^2 xt^2\,dt = \frac{8}{3}x.$$

There are historical reasons and, as we shall see later, also practical reasons for this notation. It is due to Leibniz and was rapidly adopted by Continental mathematicians. Sir Isaac never used it. The use of capitalization for primitives, as in $G' = g$, *is* due to Newton and appears in his *Principia*.

But note that "dx" and "dt" have, by themselves, no meaning at all, and that there is no multiplication involved in "$f(x)\,dx$." When integrating an abstract function f, as will happen in the proofs of some elementary theorems we will, for clarity's sake, shed the variable and write $\int_a^b f$.

With respect to the Fundamental Theorem of Calculus, as we have mentioned, both Barrow and Gregory had known of this result in special cases (and in a more geometric setting). But, as so often happens with Calculus, precursor results are found much further back. Nicole Oresme—whom we have met in Chap. 2—proved in 1361 (and perhaps before him, the Merton scholars at Oxford), the *Merton acceleration theorem* regarding uniformly accelerated movement: the distance $f(t)$ covered by an object moving with linearly increasing velocity $v(t)$

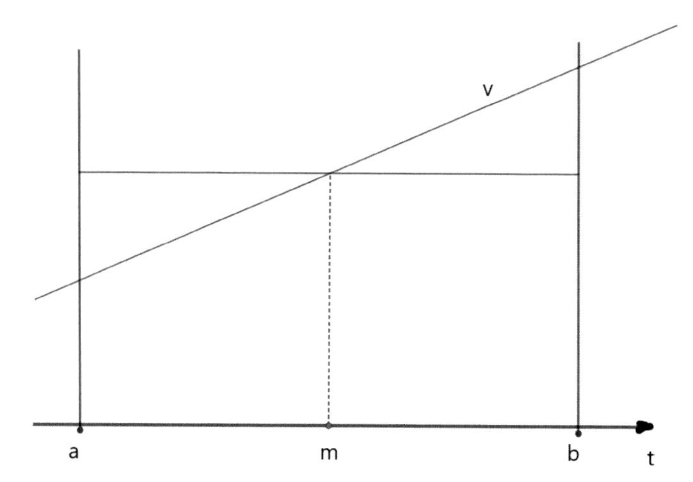

would be $(b - a)$ times the velocity at the midpoint m. That is, in our notation,

$$f(b) - f(a) = (b - a)v(m) = \int_a^b v(t)\,dt.$$

Since $f' = v$, this is Barrow's rule in a very particular case. We have mentioned above that in the first half of the XVIIth Century Bonaventura Cavalieri heuristically conceived areas as a "sum of lines." As we will see in Chap. 8, this also put him close to the Fundamental Theorem of Calculus.

It is hard to exaggerate the importance of the Fundamental Theorem of Calculus, both from the practical point of view—for it provides a method to effectively calculate innumerable integrals—and from the theoretical point of view, for it is the nexus between the two most important notions in Calculus: the derivative and the integral. The procedure we used to prove the theorem is also important, and you will see that we will repeat it in several occasions: when we talk about length of curves, volumes, surfaces, and center of mass. Like Cavalieri, we may consider the area under the graph of a non-negative continuous function $f : [a, b] \longrightarrow \mathbb{R}$ as a sum of vertical segments of length $f(x)$:

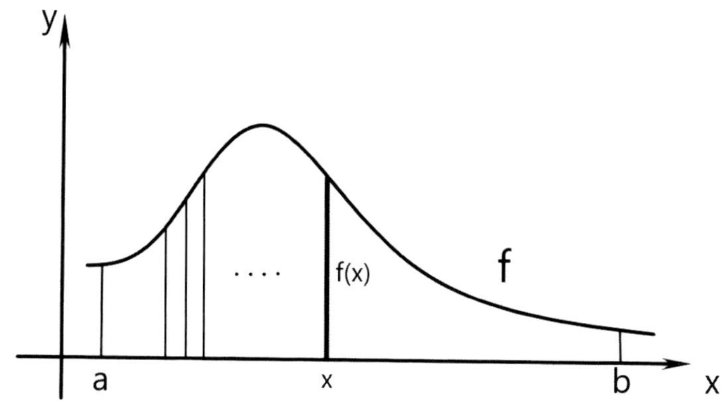

One might ask if the same area could be considered a sum of horizontal segments.

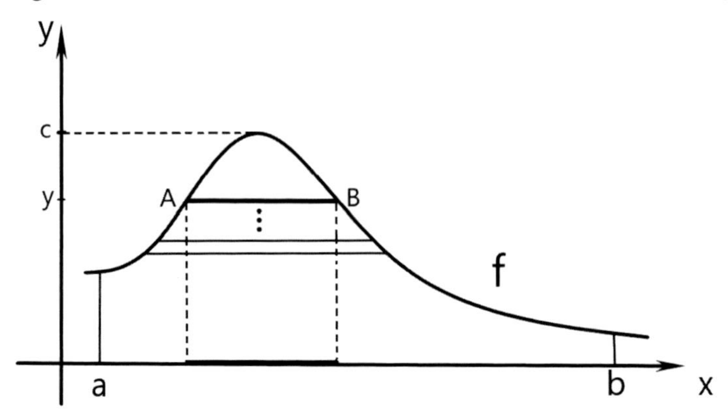

The answer is yes and that it may be useful to do so. But let's see what the length of the segment $\overline{AB}$ is: it's the same as that of its projection on the x-axis, the set $\{x \in [a, b] : f(x) \geq y\}$. If we call $m\{f(x) \geq y\}$ the measure of this segment, the area under the curve may be written as

$$\int_0^c m\{f(x) \geq y\}\, dy,$$

where c is the maximum value of f on $[a, b]$.

We may justify this a little better by a procedure analogous to the one we used to prove the Fundamental Theorem of Calculus. Call $A(y)$ the area under the graph from height 0 to height y:

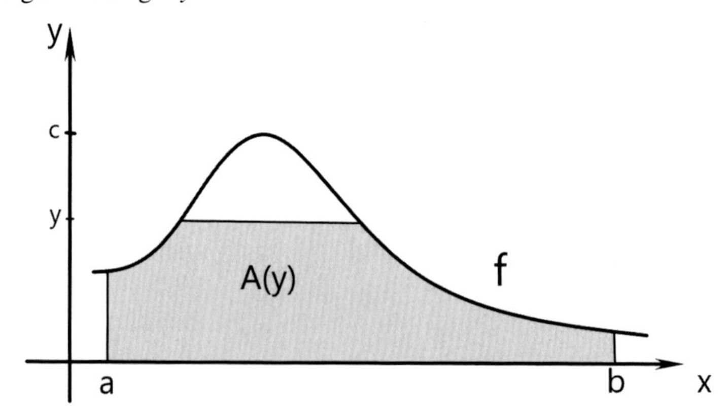

Thus, $A(y + h) - A(y)$ is the area of the horizontal strip

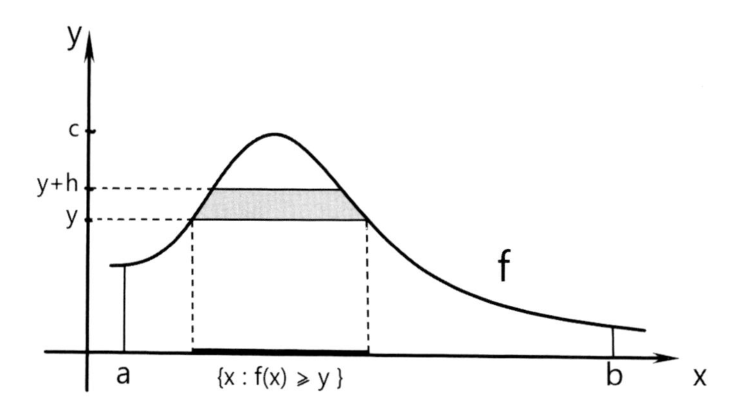

Now, call m_h the smallest horizontal length within that strip and M_h the largest. We then have

$$m_h\, h \leq A(y+h) - A(y) \leq M_h\, h$$

and

$$m_h \leq \frac{A(y+h) - A(y)}{h} \leq M_h.$$

Having h tend to zero,

$$m\{f(x) \geq y\} \leq \lim_{h\to 0} \frac{A(y+h) - A(y)}{h} \leq m\{f(x) \geq y\}.$$

Then $A'(y) = m\{f(x) \geq y\}$, and the area under the curve is

$$A(c) = A(c) - A(0) = \int_0^c A'(y)\, dy = \int_0^c m\{f(x) \geq y\}\, dy.$$

We may then write:

$$\int_a^b f(x)\, dx = \int_0^c m\{f(x) \geq y\}\, dy.$$

This is sometimes called *the layer-cake representation.*

Buffon's Needle

Georges Louis Leclerc, Count of Buffon (1707–1788) liked to have a drink with his friends. Between one glass and the next, questions came up such as: How much would you bet that if I throw this needle into the air it will fall intersecting a groove between two boards of this table? To bet intelligently, the Count needed to know the probabilities. Let's analyze the problem: we have a table made with boards of a certain width (to fix a measure, say that this width is 1). We drop on the table a needle of length $\ell \leq 1$, (ℓ is the length of the needle measured in the unit "width of a board") and see how it falls:

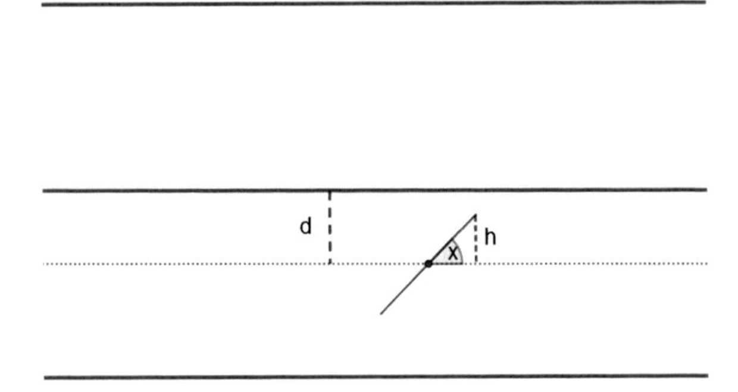

Does it intersect a groove or not? This depends on two things:

(i) The distance d from the center of the needle to the closest groove, and
(ii) The angle x of the needle with respect to the horizontal.

If the vertical distance h is larger or equal to d, the needle will intersect a groove. This distance h is $\frac{\ell}{2}\sin x$. Therefore the needle will touch a groove if $d \leq \frac{\ell}{2}\sin x$. Now, d will have values between 0 and $\frac{1}{2}$, while x will be between 0 and π (parallel to a groove and, after half a turn, again parallel). Thus, each possible position of the needle corresponds to a point in the rectangle $[0, \pi] \times [0, \frac{1}{2}]$:

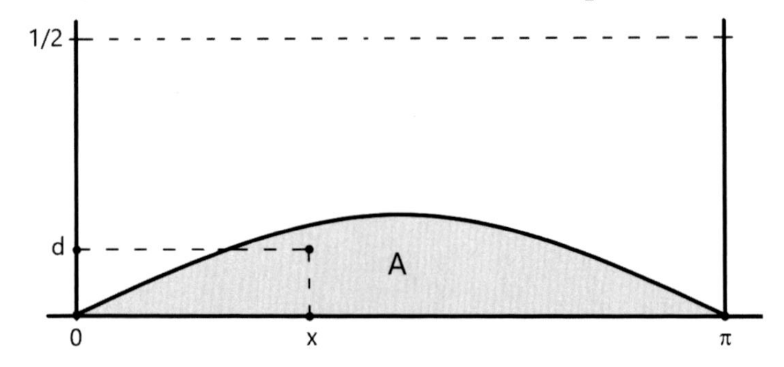

And the needle will touch a groove if $d \leq \frac{\ell}{2}\sin x$, in other words $(x, d) \in A$, and if not, it won't. Therefore the probability of the falling needle intersecting a groove is the ratio

$$\frac{\text{area}(A)}{\text{area of the rectangle}}.$$

The area of the rectangle is $\frac{\pi}{2}$; and the area of A is:

$$\begin{aligned}\int_0^\pi \frac{\ell}{2}\sin x\,dx &= \frac{-\ell}{2}\cos x\Big|_0^\pi\\ &= \frac{-\ell}{2}\cos\pi + \frac{\ell}{2}\cos 0\\ &= \frac{\ell}{2}+\frac{\ell}{2}\\ &= \ell.\end{aligned}$$

Then the probability of the needle touching a groove is $\frac{2\ell}{\pi}$. Now the good Count knows the odds.

Irrationality of π

We have seen that $\sqrt{2}$ and e are irrational numbers. Let's see now that π is also irrational. This was proved by Johann Lambert in the XVIIIth Century; the proof we give here is by the mathematician Ivan Niven (1947) [10]. We will use higher order derivatives of a function f: $f^{(2)}$ denotes f'', $f^{(3)} = f'''$,..., $f^{(k)}$ the k-th derivative of f.

The Number π Is Irrational

Suppose to the contrary, that $\pi = \frac{a}{b}$, with a, $b \in \mathbb{N}$, and we will reach a contradiction. Having fixed such a and b, we may define—for any $n \in \mathbb{N}$—the function

$$f(x) = \frac{1}{n!}\left(x(a-bx)\right)^n.$$

Note several things regarding this f:

(i) $f(0) = 0 = f(\pi)$.
(ii) $f(x) > 0$ in the interval $(0, \pi)$.
(iii) $f(x) = f(\pi - x)$. Note that $x(a - bx) = xb(\pi - x)$.
(iv) f reaches its maximum (within $(0, \pi)$) at $\frac{\pi}{2}$, where its value is $\frac{(a\pi/4)^n}{n!}$, for by the arithmetic-geometric inequality $x(\pi - x) \leq \left(\frac{\pi}{2}\right)^2$.

Also, $f^{(k)}(0)$ and $f^{(k)}(\pi)$ are whole numbers. To see this, note that f is a polynomial of degree $2n$:

$$f(x) = \frac{1}{n!}\left(c_n x^n + c_{n+1}x^{n+1} + \cdots + c_{2n}x^{2n}\right), \text{ with } c_k \in \mathbb{Z},$$

so if we differentiate f k times, we will have, for $x = 0$:

$$f^{(k)}(0) = \begin{cases} 0, & \text{if } 0 \leq k < n \\ \frac{k!}{n!} c_k, & \text{if } n \leq k \leq 2n \\ 0, & \text{if } k > 2n. \end{cases}$$

Analogously, differentiating $f(\pi - x)$, we have $f^{(k)}(\pi - x) = (-1)^k f^{(k)}(x)$, and

$$f^{(k)}(\pi) = (-1)^k f^{(k)}(0)$$

is also a whole number for any k.

Now define

$$F(x) = f(x) - f''(x) + f^{(4)}(x) - \cdots + (-1)^n f^{(2n)}(x).$$

Note that $F(\pi)+F(0)$ is a whole number, and that $F(x)+F''(x) = f(x)$. Calculate

$$\begin{aligned} [F'(x)\sin x - F(x)\cos x]' &= F''(x)\sin x + F'(x)\cos x - F'(x)\cos x + F(x)\sin x \\ &= (F''(x) + F(x))\sin x \\ &= f(x)\sin x. \end{aligned}$$

Thus, $F'(x)\sin x - F(x)\cos x$ is a primitive of $f(x)\sin x$, and then

$$\begin{aligned} 0 < \int_0^\pi f(x)\sin x\, dx &= [F'(x)\sin x - F(x)\cos x]\Big|_0^\pi \\ &= -F(\pi)(-1) + F(0)(1) \\ &= F(\pi) + F(0), \end{aligned}$$

which is a whole number. Moreover, it is a natural number, for the integral is positive. But, on the other hand,

$$\int_0^\pi f(x)\sin x\, dx < \pi f\left(\frac{\pi}{2}\right) = \pi \frac{(a\pi/4)^n}{n!} \xrightarrow[n\to\infty]{} 0,$$

and then for large n, $\int_0^\pi f(x)\sin x\, dx$ cannot be a natural number. We have reached a contradiction. Thus π is not rational. □

Improper Integrals

Until now we have considered only integrals of continuous functions defined on closed intervals and therefore, by Weierstrass' Theorem, functions which are bounded on bounded intervals. We are now interested in integrating functions which are continuous on non-closed intervals, such as $(0, 1)$, $(a, b]$ or $(0, \infty)$. Functions continuous on such intervals need not be bounded. We will call these integrals *improper integrals*. They include, for example,

$$\int_0^1 \frac{1}{\sqrt{x}}\,dx \qquad \text{and} \qquad \int_1^\infty \frac{1}{x^2}\,dx.$$

What we do in cases like these is consider the integrals on smaller closed intervals, and then take limits. Thus, for example, we will write

$$\int_0^1 \frac{1}{\sqrt{x}}\,dx = \lim_{a\to 0} \int_a^1 \frac{1}{\sqrt{x}}\,dx$$

and

$$\int_1^\infty \frac{1}{x^2}\,dx = \lim_{b\to\infty} \int_1^b \frac{1}{x^2}\,dx,$$

if these limits exist! Let's see some examples

Example Integrals of x^c.

Consider first the interval $(0, 1]$: for $c = -1$ we have

$$\int_0^1 \frac{1}{x}\,dx = \lim_{a\to 0} \int_a^1 \frac{1}{x}\,dx = \lim_{a\to 0} \ln x \Big|_a^1 = \lim_{a\to 0} (-\ln a) = \infty.$$

For $c \neq -1$,

$$\int_0^1 x^c\,dx = \lim_{a\to 0} \int_a^1 x^c\,dx = \lim_{a\to 0} \frac{x^{c+1}}{c+1}\Big|_a^1 = \lim_{a\to 0} \left(\frac{1}{c+1} - \frac{a^{c+1}}{c+1}\right)$$

$$= \begin{cases} \frac{1}{c+1}, & \text{if } c > -1 \quad (c+1 > 0) \\ \infty, & \text{if } c < -1 \quad (c+1 < 0). \end{cases}$$

On the interval $[1, \infty)$: for $c = -1$ we have

$$\int_1^\infty \frac{1}{x}\,dx = \lim_{b\to\infty} \ln x \Big|_1^b = \lim_{b\to\infty} \ln b = \infty$$

For $c \neq -1$,

$$\int_1^\infty x^c\,dx = \lim_{b\to\infty}\int_1^b x^c\,dx = \lim_{b\to\infty}\frac{x^{c+1}}{c+1}\Big|_1^b = \lim_{b\to\infty}\left(\frac{b^{c+1}}{c+1} - \frac{1}{c+1}\right)$$

$$= \begin{cases} \infty, & \text{if } c > -1 \quad (c+1 > 0) \\ \frac{-1}{c+1}, & \text{if } c < -1 \quad (c+1 < 0). \end{cases}$$

Thus, for example, the area under the graph of $\frac{1}{x}$ is infinite over $(0, 1]$ and also over $[1, \infty)$; but

$$\int_0^1 \frac{1}{\sqrt{x}}\,dx = 2 \qquad \text{and} \qquad \int_1^\infty \frac{1}{x^2}\,dx = 1.$$

We will sometimes simply write (if $F' = f$), $\int_a^\infty f = F(x)\Big|_a^\infty$ instead of $\lim_{b\to\infty} F(x)\Big|_a^b$.

Let's turn to another convergence criterion for series with positive terms. This one is called the Integral Criterion.

Integral Criterion *If* $f : [1, \infty) \to \mathbb{R}$ *is continuous, positive, and decreasing,*

$$\sum_{k=1}^\infty f(k) \text{ converges} \iff \int_1^\infty f < \infty.$$

To see why we need only take a look at the picture

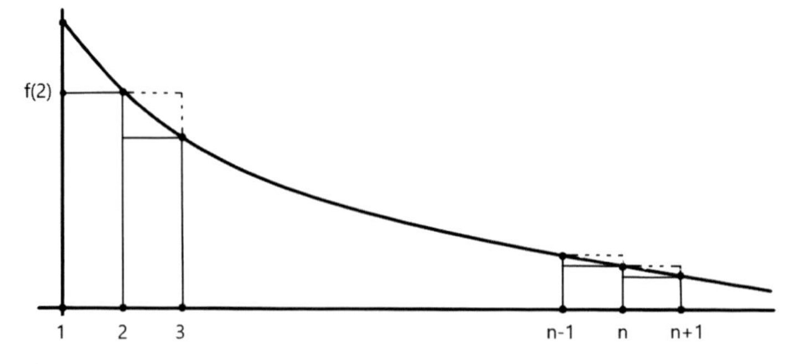

and note that

$$\int_2^{n+1} f \le \sum_{k=2}^n f(k) \le \int_1^n f.$$

□

Using this criterion we may see (for the fourth time!) that the harmonic series $\sum_{k=1}^{\infty} \frac{1}{k}$ diverges, for $\int_1^{\infty} \frac{1}{x}\,dx = \infty$. But we also see that $\sum_{k=1}^{\infty} \frac{1}{k^2}$ converges, for $\int_1^{\infty} \frac{1}{x^2}\,dx \leq \infty$. In fact, $\sum_{k=1}^{\infty} \frac{1}{k^\alpha}$ converges for any $\alpha > 1$.

The same ideas will help us understand just how rapidly the partial sums of the harmonic series tend to infinity. If in the above picture we put $f(x) = \frac{1}{x}$, we obtain the inequalities

$$\int_2^{n+1} \frac{1}{x}\,dx < \sum_{k=2}^{n} \frac{1}{k} < \int_1^{n} \frac{1}{x}\,dx.$$

Thus,

$$\ln(n+1) - \ln 2 < \sum_{k=2}^{n} \frac{1}{k} < \ln n,$$

and adding 1,

$$\ln(n+1) + 1 - \ln 2 < \sum_{k=1}^{n} \frac{1}{k} < \ln n + 1.$$

Now divide everything by $\ln n$,

$$\frac{\ln(n+1)}{\ln n} + \frac{1-\ln 2}{\ln n} < \left(\frac{\sum_{k=1}^{n} \frac{1}{k}}{\ln n}\right) < 1 + \frac{1}{\ln n}.$$

Both the left-hand side and the right-hand side tend to 1. Therefore, given any $\varepsilon > 0$ we have, for sufficiently large n,

$$1 - \varepsilon < \left(\frac{\sum_{k=1}^{n} \frac{1}{k}}{\ln n}\right) < 1 + \varepsilon.$$

Then, for large n,

$$(1-\varepsilon)\ln n < \sum_{k=1}^{n} \frac{1}{k} < (1+\varepsilon)\ln n,$$

so the harmonic series grows like $\ln n$.

If we define $x_n = \sum_{k=1}^{\infty} \frac{1}{k} - \ln(n+1)$, it is easy to check by looking at the picture above that the sequence (x_n) is increasing and bounded, thus convergent. Its limit is called γ, Euler's constant.

Integration and Sums: *Linearity of the Integral*

Having established the relationship between integral and derivative through the Fundamental Theorem of Calculus, some of the things we know about derivatives translate into results regarding integrals. This is what will happen with the relation between integration and sums, integration and products, and integration and composition of functions. We begin with integration and sums.

If c and d are constants, and f and g continuous functions,

$$\int_a^b (cf + dg) = c\int_a^b f + d\int_a^b g.$$

To see this: if F is a primitive of f and G a primitive of g, $(cF + dG)$ is a primitive of $cf + dg$: $(cF + dG)' = cF' + dG' = cf + dg$. Then

$$\begin{aligned}\int_a^b (cf + dg) = (cF + dG)\Big|_a^b &= cF(b) + dG(b) - cF(a) - dG(a)\\ &= c(F(b) - F(a)) + d(G(b) - G(a))\\ &= c\int_a^b f + d\int_a^b g.\end{aligned}$$

□

And we also have the following.

Mean Value Theorem for Integrals $\int_a^b f = f(c)(b - a)$ *for some* $c \in (a, b)$.

Indeed, if F is a primitive of f,

$$\int_a^b f = F(b) - F(a) = F'(c)(b - a) = f(c)(b - a)$$

for some $c \in (a, b)$, by Lagrange's Mean Value Theorem. □

Uniform Convergence—The Weierstrass *M*-Test

In Chap. 2, we have seen what it means for a sequence of numbers to converge. We said $x_n \longrightarrow a$ (or $\lim_n x_n = a$) if the following happens: for all $\varepsilon > 0$, there is a natural number n_ε such that if $n \geq n_\varepsilon$, the distance between x_n and a is smaller than ε:

$$|x_n - a| < \varepsilon.$$

We now ask ourselves how we could define the convergence of a sequence of functions f_n to another function f. For each x, the sequence $f_n(x)$ is a sequence of numbers, so we will want $f_n(x) \longrightarrow f(x)$. We could use any of the following two options, (A) or (B):

(A) For all x and all $\varepsilon > 0$, there is an n_ε such that, if $n \geq n_\varepsilon$, $|f_n(x) - f(x)| < \varepsilon$.
(B) For all $\varepsilon > 0$, there is an n_ε such that, if $n \geq n_\varepsilon$, $|f_n(x) - f(x)| < \varepsilon$ for all x.

As you can see, the only difference between these definitions is that in the first "for all x" is at the beginning, and in the second, it is at the end. It is a big difference. In (A), given x and ε we ask for the existence of n_ε. This n_ε depends on x and on ε. On the other hand, in (B), we ask for the existence of n_ε given only ε, and the final inequality must hold for all x. In other words: in (A), the n_ε depends on ε and on x, while in (B) the n_ε depends only on ε. This is stronger, for the same n_ε works for every x.

The first form of convergence, (A), is called *pointwise convergence*; the second, (B), is called *uniform convergence*. As we have mentioned, if $f_n \longrightarrow f$ uniformly, then $f_n \longrightarrow f$ pointwise also. But the converse is not true. Let's see an example:

Example Define $f_n : [0, 1] \longrightarrow \mathbb{R}$ thus: $f_n(x) = x^n$, and set

$$f(x) = \begin{cases} 0, & \text{if } x < 1 \\ 1, & \text{if } x = 1. \end{cases}$$

Then $f_n \longrightarrow f$ pointwise, but not uniformly. Let's see: take $\varepsilon > 0$. We wish to see for which values of n we will have

$$|f_n(x) - f(x)| < \varepsilon.$$

On $x = 1$ all the f_n's and f are 1, so the difference here is zero. For $x < 1$, the inequality

$$|f_n(x) - f(x)| = x^n < \varepsilon$$

is, applying logarithm,

$$n \ln x < \ln \varepsilon$$

$$n > \frac{\ln \varepsilon}{\ln x} \quad \text{(note that } \ln x \text{ is negative!).}$$

Thus, the value of n from which $|f_n(x) - f(x)| < \varepsilon$ depends on ε and on x... but notice how it depends on x: as x tends to 1, $\ln x$ is closer to zero, and the quotient

$\frac{\ln \varepsilon}{\ln x}$ tends to infinity... there will be no value of n which works for all x. $f_n \longrightarrow f$ pointwise, but not uniformly.

Note something else: the functions f_n in this example are all continuous, but the limit function f is not. However, we have the following.

Proposition *If f_n are continuous, and $f_n \longrightarrow f$ uniformly, the limit function f will be continuous.*

Let's see a proof of this: Say we have f_n continuous, converging uniformly to f. We prove the continuity of f at any point s. If $f(s) < v$, take $\varepsilon < \frac{v-f(s)}{2}$ (so that $f(s) + 2\varepsilon < v$). For this ε there is an n such that $|f_n(x) - f(x)| < \varepsilon$ for all x. In particular, $f_n(s) < f(s) + \varepsilon$; and as f_n is continuous, for all x in a neighborhood of s we will also have $f_n(x) < f(s) + \varepsilon$. Now,

$$\begin{aligned} f(x) &= (f(x) - f_n(x)) + f_n(x) \\ &< \varepsilon + f_n(x) \\ &< \varepsilon + f(s) + \varepsilon \\ &= f(s) + 2\varepsilon \\ &< v. \end{aligned}$$

This happens for every x in a neighborhood of s. Analogously, if $f(s) > u$, we will also have $f(x) > u$ for all x in a neighborhood of s. f is continuous. □

Suppose that f_n are continuous, and $f_n \longrightarrow f$ pointwise. The example above shows that f may *not* be continuous at a, so $\lim_{x \longrightarrow a} f(x) \neq f(a)$. Then

$$\lim_{x \longrightarrow a} \lim_{n \longrightarrow \infty} f_n(x) = \lim_{x \longrightarrow a} f(x) \neq f(a) = \lim_{n \longrightarrow \infty} f_n(a) = \lim_{n \longrightarrow \infty} \lim_{x \longrightarrow a} f_n(x).$$

This shows us something very important that you must not forget:

Limits do NOT commute

and there are so many limit processes in analysis: summing a series, differentiating, integrating... the commutation of these processes is to be handled with extreme care. The theorem we have just seen relating uniform convergence and continuity shows that uniform convergence permits the commutation of some limits. We will see later on that something similar happens with uniform convergence and integration. The following is a criterion which will help us sum a series of functions.

The Weierstrass M-Test *Given functions $f_k : B \longrightarrow \mathbb{R}$, if there are positive numbers M_k such that $\sup_{x \in B} |f_k(x)| \leq M_k$ for all k, and $\sum_{k=1}^{\infty} M_k$ converges, then $\sum_{k=1}^{\infty} f_k(x)$ converges absolutely and uniformly on B.*

Let's prove this. For each $x \in B$ the series $\sum_{k=1}^{\infty} f_k(x)$ converges absolutely by the comparison criterion: $|f_k(x)| \leq M_k$ and $\sum_{k=1}^{\infty} M_k$ converges. Write then $f(x) = \sum_{k=1}^{\infty} f_k(x)$. We want to see that the partial sums $\sum_{k=1}^{n} f_k$ of the series converge uniformly to f. Given $\varepsilon > 0$ take n_ε so large that $\sum_{k>n_\varepsilon} M_k < \varepsilon$. We then have, for any $x \in B$,

$$\left| f(x) - \sum_{k=1}^{n_\varepsilon} f_k(x) \right| = \left| \sum_{k>n_\varepsilon} f_k(x) \right| \leq \sum_{k>n_\varepsilon} |f_k(x)| \leq \sum_{k>n_\varepsilon} M_k < \varepsilon.$$

Thus, $\sum_{k=1}^{n} f_k$ converges uniformly to f as n grows. □

Note then that if the functions f_k are continuous, so is f. One last example:

Example Geometric series on $[-b, b]$, with $b < 1$.

Consider $f_k : [-b, b] \longrightarrow \mathbb{R}$ such that $f_k(x) = x^k$. Then $\sup_{[-b,b]} |x^k| \leq b^k$, and $\sum_{k=1}^{\infty} b^k = \frac{b}{1-b} < \infty$. Thus,

$$\sum_{k=0}^{n} x^k \longrightarrow \frac{1}{1-x}$$

absolutely and uniformly on $[-b, b]$, by the Weierstrass M-test.

Karl Weierstrass (1815–1897) was a German mathematician who gave important results on uniform convergence and on the foundations of mathematics. He was also one of the founders of the calculus of variations.

We mention now a result on the integration of a limit of functions: the integral of a uniform limit of functions will be the limit of the integrals of these functions. Recall that we integrate only continuous functions. There are more general versions of this theorem for Riemann integrable or Lebesgue integrable functions. We will also see, by an example, that if the convergence is not uniform, this result may not hold. First, the theorem:

Uniform Convergence and Integration *If the continuous functions f_n converge uniformly to f on $[a, b]$, then*

$$\int_a^b f = \lim_{n\to\infty} \int_a^b f_n.$$

Let's see: note that the function f is continuous, and we have

$$\left| \int_a^b f_n - \int_a^b f \right| = \left| \int_a^b (f_n - f) \right| \leq \int_a^b |f_n - f|, \qquad (*)$$

but given $\varepsilon > 0$ there exists n_ε such that for all larger n, one has

$$|f_n(x) - f(x)| < \varepsilon,$$

This *for all* $x \in [a, b]$. Thus,

$$(*) < \int_a^b \varepsilon = \varepsilon(b-a),$$

which may be made as small as required. □

Let's see what may happen if the convergence is only pointwise.

Example Take $f_n : (0, 1] \to \mathbb{R}$ such that $f_n(x) = \frac{x^{\frac{1}{n}-1}}{n}$. Note that since the exponent is larger than -1, the (improper) integrals of these functions exist over $(0, 1]$. We have, in fact

$$\begin{aligned}\int_0^1 f_n &= \lim_{a\to 0} \int_a^1 \frac{x^{\frac{1}{n}-1}}{n}\,dx \\ &= \lim_{a\to 0} x^{\frac{1}{n}}|_a^1 \\ &= \lim_{a\to 0} (\sqrt[n]{1} - \sqrt[n]{a}) \\ &= 1,\end{aligned}$$

so $\lim_{n\to\infty} \int_0^1 f_n = 1$. But the functions f_n converge pointwise to zero on $(0, 1]$, so

$$\int_0^1 \lim_{n\to\infty} f_n = \int_0^1 0 = 0.$$

What happens is that the convergence of the f_n to zero in this example is not uniform:

$$\begin{aligned}\text{for } \frac{x^{\frac{1}{n}-1}}{n} &< \varepsilon, \text{ we must have} \\ x^{\frac{1}{n}-1} &< n\varepsilon \\ \left(\frac{1}{n} - 1\right) \ln x &< \ln(n\varepsilon) \\ \ln x &> \frac{\ln(n\varepsilon)}{(\frac{1}{n} - 1)} = -\frac{\ln(n\varepsilon)}{(1 - \frac{1}{n})},\end{aligned}$$

but for x close to zero, since $\ln x \to -\infty$, we need larger and larger n for this inequality to hold. There is no n_ε that works for all x.

Gregory's Series

James Gregory (1638–1675), Scottish mathematician, found the Taylor series of various trigonometric functions, and also gave an early version of the Fundamental Theorem of Calculus.

We saw in Chap. 2 that Gregory's series, $\sum_{k=0}^{\infty} \frac{(-1)^k}{2k+1}$ converges, by the Leibniz criterion for alternating series. We'll now see what it converges to, as an application of the results we have seen on uniform convergence. Gregory proved this result in 1668. It was, however, known to the Hindu mathematician Madhava of Sangamagrama in the XIVth Century.

Sum of Gregory's Series $\sum_{k=0}^{\infty} \frac{(-1)^k}{2k+1} = \frac{\pi}{4}$.

Recall that $\arctan t$ is a primitive of $\frac{1}{1+t^2}$. Thus,

$$\begin{aligned}
\frac{\pi}{4} &= \arctan 1 = \lim_{b\to 1} \arctan b \\
&= \lim_{b\to 1} \int_0^b \frac{1}{1+t^2}\, dt \\
&= \lim_{b\to 1} \int_0^b \frac{1}{1-(-t^2)}\, dt \text{ , which, written as a geometric series is} \\
&= \lim_{b\to 1} \int_0^b \sum_{k=0}^{\infty} (-t^2)^k\, dt \\
&= \lim_{b\to 1} \int_0^b \sum_{k=0}^{\infty} (-1)^k t^{2k}\, dt \\
&= \lim_{b\to 1} \sum_{k=0}^{\infty} (-1)^k \int_0^b t^{2k}\, dt, \text{ for on } [-b, b] \text{ the convergence is uniform} \\
&= \lim_{b\to 1} \sum_{k=0}^{\infty} (-1)^k \frac{t^{2k+1}}{2k+1}\Big|_0^b \\
&= \lim_{b\to 1} \sum_{k=0}^{\infty} (-1)^k \frac{b^{2k+1}}{2k+1}. \qquad (*)
\end{aligned}$$

Call $f(b) = \sum_{k=0}^{\infty} (-1)^k \frac{b^{2k+1}}{2k+1}$. In order to commute the limit and the series in $(*)$, we need to prove that f is a continuous function on $[0, 1]$. It is there a uniform limit of continuous functions: if we set $f_n(x) = \sum_{k=0}^{n} (-1)^k \frac{x^{2k+1}}{2k+1}$, we have

$$|f(x) - f_n(x)| = \left|\sum_{k>n}(-1)^k \frac{x^{2k+1}}{2k+1}\right|,$$

but, being an alternating series, its tail is smaller than the last term added

$$\left|\frac{x^{2n+3}}{2n+3} - \frac{x^{2n+1}}{2n+1}\right| < \frac{x^{2n+1}}{2n+1} \leq \frac{1}{2n+1} < \varepsilon$$

if $n > n_\varepsilon$ (independently of x). Then

$$(*) = \lim_{b\to 1} f(b) = f(1) = \sum_{k=0}^{\infty} \frac{(-1)^k}{2k+1}.$$

Therefore,

$$\sum_{k=0}^{\infty} \frac{(-1)^k}{2k+1} = \frac{\pi}{4}.$$

□

Note that in the proof we have commuted an integral and an infinite series. We have been able to do this because, as we saw when discussing the Weierstrass M-test, the geometric series converges uniformly for $b < 1$.

Integration and Products: *Integration by Parts*

From the formula for the derivative of a product, we immediately obtain a very useful integration method, called *integration by parts.*

Integration by Parts *If F is a primitive of f, and G is a primitive of g,*

$$\int_a^b fG = F(x)G(x)\Big|_a^b - \int_a^b Fg.$$

Indeed, since $(FG)' = F'G + FG' = fG + Fg$,

$$F(x)G(x)\Big|_a^b = \int_a^b (FG)' = \int_a^b fG + \int_a^b Fg, \quad \text{from where}$$

$$\int_a^b fG = F(x)G(x)\Big|_a^b - \int_a^b Fg.$$

□

Let's see a couple of examples.

Example An integral formula for $n!$:

$$n! = \int_0^\infty e^{-t} t^n \, dt.$$

We calculate the integral using parts n times,

$$\begin{aligned}
\int_0^\infty e^{-t} t^n \, dt &= -e^{-t} t^n \Big|_0^\infty - \int_0^\infty -e^{-t} n t^{n-1} \, dt \\
&= 0 - 0 + n \int_0^\infty e^{-t} t^{n-1} \, dt = \cdots \\
\cdots &= n! \int_0^\infty e^{-t} \, dt \\
&= n! (-e^{-t}) \Big|_0^\infty \\
&= n!
\end{aligned}$$

Example A primitive for $\ln x$.

Consider $f = 1$, and $G = \ln x$, so (setting $F = x$, $g = \frac{1}{x}$):

$$\int_a^b \ln x \, dx = x \ln x \Big|_a^b - \int_a^b 1 \, dx = (x \ln x - x) \Big|_a^b.$$

In other words, a primitive of $\ln x$ is $(x \ln x - x)$.

Stirling's Formula

An interesting application is (a weak form of) Stirling's Formula. If you take a calculator and calculate $n!$ for larger and larger values of n, you will soon find the limits of your calculator: $n!$ grows extremely fast. But, how fast? Using an argument similar to that which we have seen when talking about the "Integral Criterion," we write

$$\begin{aligned}
\ln(n!) = \ln(1 \cdot 2 \cdot 3 \cdots n) &= \sum_{k=1}^{n} \ln k \approx \int_1^n \ln x \, dx \\
&= (x \ln x - x) \Big|_1^n = n \ln n - n + 1 \\
&\approx n \ln n - n,
\end{aligned}$$

and applying the exponential function,

$$n! \approx \mathrm{e}^{n \ln n - n} = \frac{n^n}{\mathrm{e}^n},$$

a useful formula which we will turn to later on. Actually, Stirling proved something more precise regarding the growth of $n!$:

$$n! \approx \frac{n^n}{e^n}\sqrt{2\pi n}$$

which we will see in Chap. 8.

Integration and Composition: *Integration by Substitution*

The formulas for derivative of a sum and for derivative of a product had important consequences when we related them—through the Fundamental Theorem of Calculus—with the integral. The first gave us the linearity of the integral, while the second gave us the method of integration by parts. We will now see what consequence the chain rule has for integration. Recall that the composition of functions lets us think of a function as depending on one or another variable, in the following sense:

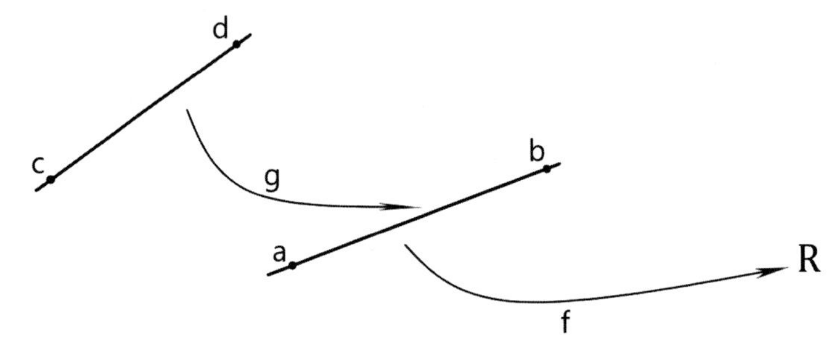

if we think that $t \in [c, d]$ and $x = g(t) \in [a, b]$, the function f may be considered as depending on the variable x writing $f(x)$, or the variable t, writing $f(g(t))$; where $x = g(t)$ expresses how x varies while moving t. The integral

$$\int_a^b f$$

depends only on the function f and the interval $[a, b]$. But if we wish to integrate f considering that its variable moves in the interval $[c, d]$, we must apply g to transform this interval into $[a, b]$. Here the function g' has a role to play, for it measures the velocity with which g changes the variable t for the variable x. But let's see what the theorem says.

Change of Variables Theorem *If $g : [c, d] \to [a, b]$ is such that $g(c) = a$, $g(d) = b$, and g' is continuous, then for any continuous $f : [a, b] \to \mathbb{R}$,*

$$\int_a^b f = \int_c^d (f \circ g)g'.$$

It is easy to see why. Say F is a primitive of f. Then, by the chain rule, $(F \circ g)$ is a primitive of $(f \circ g)g'$: indeed, $(F \circ g)'(t) = f(g(t))g'(t)$. So

$$\int_a^b f = F(b) - F(a) = F(g(d)) - F(g(c)) = \int_c^d (f \circ g)g'.$$

□

A Note on Notation

There is notation regarding the Change of Variables Theorem intended to help in its use and application. It is, for example, common practice to use $x(t)$ instead of $g(t)$, and to write the theorem in the form

$$\int_a^b f(x)\, dx = \int_c^d f(x(t))x'(t)\, dt.$$

The notation "dx" and "dt" is indicative of how we name the variable in the interval over which we are integrating. By themselves, "dx" and "dt" do not have any meaning. Beyond its theoretical importance the Change of Variables Theorem provides a method for calculating integrals known as integration by "substitution," in which a complicated expression is substituted by a simpler one. For instance,

Example Calculate

$$\int_0^2 t\mathrm{e}^{t^2}\, dt.$$

We proceed to "substitute" x for t^2 (because we know a primitive for e^x but not one for e^{t^2}:

$$x = x(t) = t^2$$

differentiating,

$$x'(t) = 2t.$$

Also, since $x(0) = 0$ and $x(2) = 4$, we may finally write

$$\begin{aligned}\int_0^2 te^{t^2}\,dt &= \int_0^2 \frac{x'(t)}{2}\mathrm{e}^{x(t)}\,dt\\ &= \frac{1}{2}\int_0^4 \mathrm{e}^x\,dx = \frac{1}{2}\mathrm{e}^x\Big|_0^4 = \frac{1}{2}(\mathrm{e}^4 - 1).\end{aligned}$$

The dt and dx notation serves nicely to mechanize the procedure:

$$\begin{aligned}x &= t^2\\ \frac{dx}{dt} &= x'(t) = 2t\\ dx &= 2t\,dt,\end{aligned}$$

then

$$\int_0^2 te^{t^2}dt = \frac{1}{2}\int_0^2 e^{t^2}2t\,dt = \frac{1}{2}\int_0^4 e^x\,dx \ldots$$

Note, however, that $\frac{dx}{dt} = x'(t)$ and the "algebraic" manipulation $dx = x'(t)\,dt$ are simply useful mnemonic devices reflecting the result of the Change of Variables Theorem.

Length of Curves. The Catenary

We will see now how to define the length of a curve on the plane (parametrized by a differentiable function). Let's start with a curve on the plane which is parametrized by $p : [a, b] \to \mathbb{R}^2$, where $p(t) = (x(t), y(t))$:

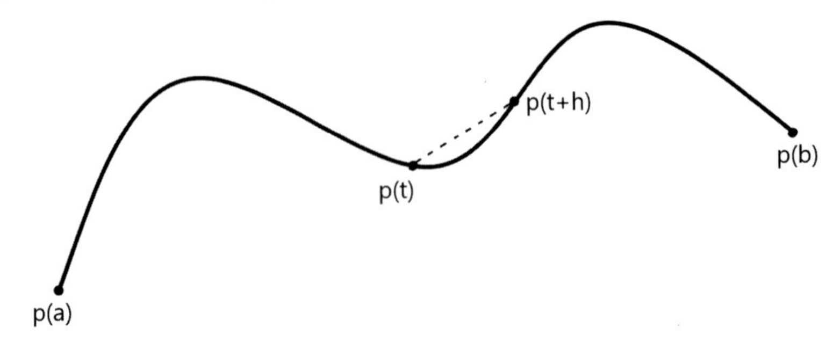

If we call $L(t)$ the length of the curve from $p(a)$ to $p(t)$, $L(t+h) - L(t)$ is the length along the curve from $p(t)$ to $p(t+h)$ which, when h is small, is like the distance between $p(t)$ and $p(t+h)$:

$$L(t+h) - L(t) \approx \sqrt{(x(t+h) - x(t))^2 + (y(t+h) - y(t))^2}.$$

If we divide by h,

$$\frac{L(t+h) - L(t)}{h} \approx \sqrt{\left(\frac{x(t+h) - x(t)}{h}\right)^2 + \left(\frac{y(t+h) - y(t)}{h}\right)^2},$$

and now have h tend to zero,

$$L'(t) = \sqrt{x'(t)^2 + y'(t)^2}.$$

Thus the length of the curve is

$$L(b) = L(b) - L(a) = \int_a^b L'(t)\,dt = \int_a^b \sqrt{x'(t)^2 + y'(t)^2}\,dt.$$

Example Length of the cycloid.

Recall that we had parametrized the cycloid by setting

$$p : [0, 2\pi] \to \mathbb{R}^2 \text{ such that } p(t) = (rt - r\sin t, r\cos t - r).$$

Hence,

$$\begin{aligned} x'(t)^2 &= (r - r\cos t)^2 = r^2 - 2r^2\cos t + r^2\cos^2 t \\ y'(t)^2 &= (-r\sin t)^2 = r^2\sin^2 t, \end{aligned}$$

so that $x'(t)^2 + y'(t)^2 = 2r^2(1 - \cos t) = 2r^2 2\sin^2\frac{t}{2}$ (using formula for the cosine of double angles: $1 - \cos 2\alpha = 2\sin^2\alpha$). Then the length of the cycloid is

$$\begin{aligned} \int_0^{2\pi} \sqrt{4r^2\sin^2\frac{t}{2}}\,dt &= \int_0^{2\pi} 2r\sin\frac{t}{2}\,dt \\ \text{with the change of variables } u = \frac{t}{2}: \quad &= 2r\int_0^{\pi} \sin u \; 2du \\ &= 4r\,(-\cos u)\Big|_0^{\pi} \\ &= 8r. \end{aligned}$$

Note that if a curve is the graph of a function f, we may parametrize it by setting $p : [a, b] \longrightarrow \mathbb{R}^2$, $p(t) = (t, f(t))$, and its length over a segment $[a, b]$ is then

$$\int_a^b \sqrt{1 + f'(t)^2}\, dt.$$

If we hang a chain from its two ends, we will see that the chain, under its own weight, forms a curve, a kind of "U," called a *catenary*. We want to determine exactly what kind of curve the catenary is. For this, suppose the catenary is the graph of a function f as in the drawing

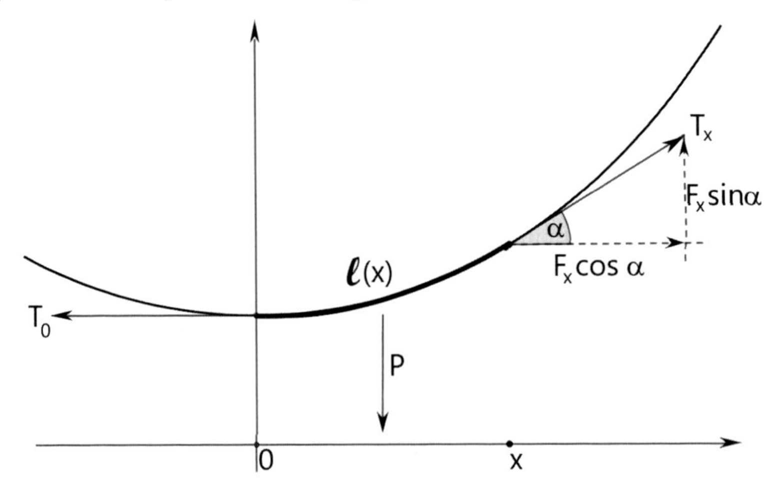

Given any x, the segment of chain which is above the interval $[0, x]$ is subject to three forces which are in equilibrium: the tensions T_0 at $(0, f(0))$ and T_x at $(x, f(x))$; and the weight P of the segment of chain between 0 and x, which is proportional to the length of that piece: $\rho\ell(x)$ (where $\ell(x)$ is the length of chain from $(0, f(0))$ to $(x, f(x))$). Bearing in mind that T_0 is horizontal, P is vertical, and T_x may be considered the sum of a horizontal component, and a vertical component, we have

$$\begin{aligned} T_0 &= (-F_0, 0) \\ P &= (0, -\rho\ell(x)) \\ T_x &= (F_x \cos\alpha,\ F_x \sin\alpha). \end{aligned}$$

Since the chain is balanced, the forces are canceled and we have

$$\begin{aligned} F_0 &= F_x \cos\alpha \\ \rho\ell(x) &= F_x \sin\alpha. \end{aligned}$$

Then,

$$f'(x) = \text{ the slope of the graph at } (x, f(x)) = \frac{F_x \sin\alpha}{F_x \cos\alpha} = \frac{\rho\ell(x)}{F_0}.$$

Putting $a = \frac{F_0}{\rho}$, we have

$$af'(x) = \ell(x) = \int_0^x \sqrt{1 + f'(t)^2}\, dt.$$

If we now differentiate, by the Fundamental Theorem of Calculus, we obtain

$$af''(x) = \sqrt{1 + f'(x)^2},$$

and squaring,

$$\begin{aligned} a^2 f''(x)^2 &= 1 + f'(x)^2 \\ a^2 f''(x)^2 - f'(x)^2 &= 1. \end{aligned}$$

But this differential equation is verified by $f(x) = a\cosh\left(\frac{x}{a}\right)$. Indeed we have

$$\begin{aligned} f'(x) &= \sinh\left(\frac{x}{a}\right) \\ \text{and } f''(x) &= \frac{1}{a}\cosh\left(\frac{x}{a}\right), \end{aligned}$$

so that

$$a^2 f''(x)^2 - f'(x)^2 = \left(\cosh\left(\frac{x}{a}\right)\right)^2 - \left(\sinh\left(\frac{x}{a}\right)\right)^2 = 1,$$

as we have seen in Exercise 3 of Chap. 3. Therefore, the catenary is the graph of the hyperbolic cosine!

Area Enclosed by a Simple Closed Curve

We will give an application of the Change of Variables Theorem to the calculation of the area enclosed by a simple closed curve. Here *simple* means that the curve does not intersect itself, while *closed* refers to the fact that it starts and ends at the same point. Thus if $p : [t_0, t_1] \longrightarrow \mathbb{R}^2$ is a parametrization of the curve, $p(t_0) = p(t_1)$, and p is otherwise one-to-one.

We will suppose that our parametrization is such that $p(t)$ moves along the curve leaving the enclosed area to the left. We also suppose that this area is comprised between the graph of a function f above, and the graph of another function g below:

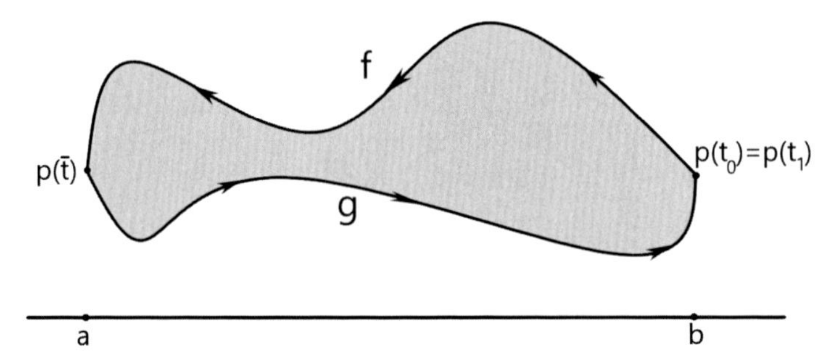

Many regions may be presented as a finite union of such areas and our results will be valid for those as well:

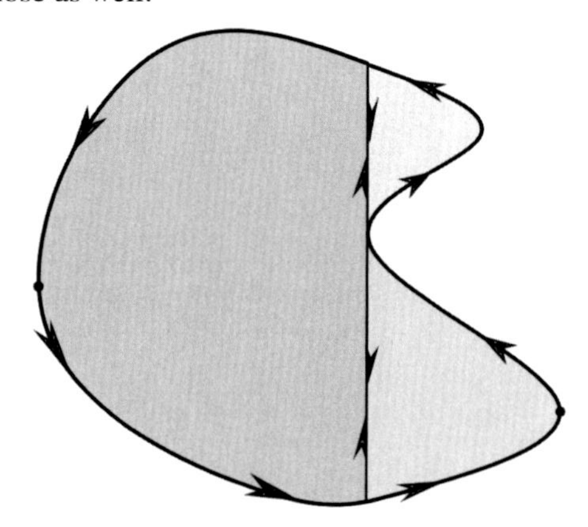

If our parametrization is $p(t) = (x(t), y(t))$, we have $x(t_0) = b$, $x(\bar{t}) = a$, and for $t_0 \leq t \leq \bar{t}$, $y(t) = f(x(t))$. Thus, if F is a primitive of f, we have, as in the Change of Variables Theorem,

$$\begin{aligned}\int_a^b f = F(b) - F(a) &= (F \circ x)(t_0) - (F \circ x)(\bar{t}) \\ &= -\left[(F \circ x)(\bar{t}) - (F \circ x)(t_0)\right] \\ &= -\int_{t_0}^{\bar{t}} (f \circ x)x' \\ &= -\int_{t_0}^{\bar{t}} yx'.\end{aligned}$$

Likewise, if G is a primitive of g, since for $\bar{t} \leq t \leq t_1$, $y(t) = g(x(t))$,

$$\int_a^b g = G(b) - G(a) = (G \circ x)(t_1) - (G \circ x)(\bar{t})$$

$$= \int_{\bar{t}}^{t_1} (g \circ x)x'$$

$$= \int_{\bar{t}}^{t_1} yx'.$$

Then the area A enclosed by the curve is

$$A = \int_a^b f - \int_a^b g = -\int_{t_0}^{\bar{t}} yx' - \int_{\bar{t}}^{t_1} yx' = -\int_{t_0}^{t_1} yx'.$$

Note that, integrating by parts, we may also write

$$A = -\int_{t_0}^{t_1} yx' = -yx\Big|_{t_0}^{t_1} + \int_{t_0}^{t_1} y'x = \int_{t_0}^{t_1} y'x$$

for since $p(t_0) = p(t_1)$, $yx\Big|_{t_0}^{t_1} = 0$. We have then

$$A = -\int_{t_0}^{t_1} yx' = \int_{t_0}^{t_1} xy'.$$

Example area of an ellipse of radii a and b.

We may parametrize its boundary by $p : [0, 2\pi] \longrightarrow \mathbb{R}^2$ with $p(t) = (a \cos t, b \sin t)$. Then,

$$A = \int_0^{2\pi} xy' = ab \int_0^{2\pi} \cos^2 t \, dt = ab\pi.$$

Exercises

1 Without integrating, give a geometric argument to justify the equality

$$\int_0^1 x^{\frac{1}{n}} \, dx + \int_0^1 x^n \, dx = 1.$$

2 Without calculating the integral, find maxima and minima of

$$F(x) = \int_0^x \sin t \, dt.$$

3 Calculate the derivatives of the following functions

(i) $\int_0^{x^2} u\,du$.
(ii) $\int_x^{x+\pi} \cos u\,du$.
(iii) $\int_0^x xe^u\,du$.

4 Without integrating, show that the following expressions are equal:

$$\int_0^x f(u)(x-u)\,du \qquad \text{and} \qquad \int_0^x \left[\int_0^u f(t)\,dt\right] du.$$

5 Without integrating, show that

$$\int_x^{ax} \frac{1}{t}\,dt$$

is a constant ($x > 0$ y $a > 1$).

6 Calculate the following improper integrals:

(i) $\int_{\frac{2}{3}}^1 \frac{1}{3x-2}\,dx$.
(ii) $\int_{-1}^1 \frac{1}{\sqrt{|x|}}\,dx$.
(iii) $\int_0^\infty x^4 e^{-x^5}\,dx$.

7 Study the convergence of the series

$$\sum_{k=1}^\infty ke^{-k^2}.$$

8 Estimate $\binom{3n}{n}$ for large values of n.

9 Estimate using Stirling's formula:

(i) $2 \cdot 4 \cdot 6 \cdots (2n)$.
(ii) $1 \cdot 3 \cdot 5 \cdots (2n-1)$.
(iii) $\lim_{n \longrightarrow \infty} \frac{1 \cdot 3 \cdot 5 \cdots (2n-1)}{2 \cdot 4 \cdot 6 \cdots (2n)}$.

10 Calculate the integrals

(i) $\int_0^\pi x \cos x \, dx$.
(ii) $\int_0^1 x e^x \, dx$.
(iii) $\int_0^\pi x^2 \sin x \, dx$.
(iv) $\int_1^2 x^2 \ln x \, dx$.
(v) $\int_0^\pi e^2 x \sin x \, dx$.

11 Calculate the integrals

(i) $\int_0^1 e^{-2x} \, dx$.
(ii) $\int_0^1 \frac{3x}{(x^2+1)^7} \, dx$.
(iii) $\int_1^4 \frac{x}{\sqrt{x+1}} \, dx$.
(iv) $\int_1^2 x e^{-3x^2} \, dx$.

12 Calculate the integrals

(i) $\int_0^{2\pi} \sin nx \cos mx \, dx$.
(ii) $\int_0^{2\pi} \cos nx \cos mx \, dx$.

13 Show that the area between the cycloid and the x-axis is three times the area of the circle that generates the cycloid.

14 Prove that the area under the graph of $\cosh x$ over the interval $[a, b]$ is the same as the length of the curve over that interval.

15 Consider a parametrization $p : [a, b] \longrightarrow \mathbb{R}^2$ of a curve $\mathcal{C}$ of length L such that $p'(t) = (x'(t), y'(t)) \neq 0$ for all $t \in (a, b)$. We define

$$s(t) = \int_a^t \sqrt{x'(u)^2 + y'(u)^2} \, du.$$

$s(t)$ is then the length of the curve from $p(a)$ to $p(t)$. Note that by the Fundamental Theorem of Calculus $s'(t) = \sqrt{x'(t)^2 + y'(t)^2} > 0$ so $s : [a, b] \longrightarrow [0, L]$ is strictly increasing, and thus has an inverse $s^{-1} : [0, L] \longrightarrow [a, b]$. The function $\overline{p} = p \circ s^{-1}$ is another parametrization of $\mathcal{C}$ known as the *arc length reparametrization*. If we write $\overline{p}(s) = (\overline{x}(s), \overline{y}(s))$, prove that

(i) For each $s \in (0, L)$, we have $\overline{x}'(s)^2 + \overline{y}'(s)^2 = 1$.
(ii) The length of $\mathcal{C}$ from $\overline{p}(0)$ to $\overline{p}(\alpha)$ is α.

16 If $A = (a_1, a_2)$, $B = (b_1, b_2)$, and $C = (c_1, c_2)$, show that the area of the triangle

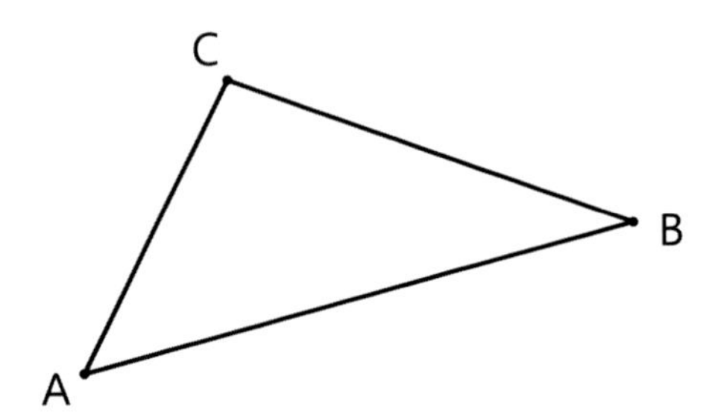

is $\mathcal{A} = \frac{1}{2}\left[a_1b_2 - a_2b_1 + b_1c_2 - b_2c_1 + a_2c_1 - a_1c_2\right]$ (parametrize, even if $p(t)$ is not differentiable everywhere...).

More Derivatives

6

The second derivative and the Taylor polynomial of order two are presented. We discuss the relationship between this polynomial and the notion of curvature. We give an application of the second order Taylor polynomial to random walk and the bell curve. Higher order Taylor polynomials and Taylor series are introduced.

Second Derivative, Best-Fitting Parabola, and Curvature

When we proved that π is irrational we used "higher order derivatives," $f^{(n)}$, of a function f. There are functions that are not so differentiable; for example, it may happen that f' exists but may not be differentiable, in which case f'' does not exist. In this chapter we will suppose that our functions are infinitely differentiable, in other words, $f', f'', \dots f^{(n)}, \dots$, exist. But now what we want to ask ourselves is: what meaning does f'' have for our function f? (Just as $f' > 0$ means f grows, and $f' < 0$ means f decreases).

Recall that when we talked about the line tangent to the graph of f at the point $(a, f(a))$,

$$y = y(x) = f(a) + f'(a)(x - a)$$

we said that this was the straight line with the "best fit" to the graph of f at that point, because

(i) $(a, f(a))$ is on the line, and
(ii) it has, at that point, the same slope as f

in other words,

I. Zalduendo, *Calculus off the Beaten Path*, SUMS Readings,
https://doi.org/10.1007/978-3-031-15765-3_6

(i) $y(a) = f(a)$.
(ii) $y'(a) = f'(a)$.

You cannot ask more of a straight line... Lines are of the form $y = C + B(x - a)$ and by imposing those two conditions i) and ii) you are fixing B and C. However, a parabola is of the form

$$y = y(x) = C + B(x - a) + A(x - a)^2$$

and therefore you can ask three things of it. Let's ask that a parabola verify

(i) $y(a) = f(a)$,
(ii) $y'(a) = f'(a)$,
(iii) $y''(a) = f''(a)$.

Bearing in mind that $y'(x) = B + 2A(x - a)$, and $y''(x) = 2A$, we have

(i) $f(a) = y(a) = C$,
(ii) $f'(a) = y'(a) = B$,
(iii) $f''(a) = y''(a) = 2A$.

Thus, the parabola

$$y(x) = f(a) + f'(a)(x - a) + \frac{1}{2}f''(a)(x - a)^2$$

is the one that best fits the graph of f near the point $(a, f(a))$. It verifies:

(i) $(a, f(a))$ is on the parabola,
(ii) the parabola has, at that point, the same slope as f, and
(iii) the parabola has, at that point, the same *curvature* as f.

Note that we have not yet defined curvature. This will be a geometric notion depending completely on $f'(a)$ and $f''(a)$, and thus determined by the best-fitting parabola, the Taylor polynomial of order two.

The Taylor Polynomial of Order Two

The degree two polynomial that we have just defined is called the *Taylor polynomial of order two*. It deserves a couple of comments:

(a) How well does the parabola $y(x)$ fit $f(x)$? Consider the difference $f(x) - y(x)$ and let's divide it by $(x - a)^2$:

$$\frac{f(x) - y(x)}{(x-a)^2}.$$

We are dividing by something very small: when x is near to a, $(x - a)$ is very small, and $(x - a)^2$ is smaller still. However, if we have $x \to a$:

$$\begin{aligned}
\frac{f(x) - y(x)}{(x-a)^2} &\longrightarrow \frac{0}{0} \quad \text{apply L'Hôpital's rule} \\
\frac{f'(x) - y'(x)}{2(x-a)} &\longrightarrow \frac{0}{0} \quad \text{again L'Hôpital} \\
\frac{f''(x) - y''(x)}{2} &\longrightarrow \frac{0}{2} = 0.
\end{aligned}$$

So when $x \to a$, the difference $f(x) - y(x)$ tends to zero, even when divided by $(x - a)^2$. We write then $f(x) - y(x) = \mathrm{o}((x - a)^2)$:

$$f(x) = f(a) + f'(a)(x-a) + \frac{1}{2} f''(a)(x-a)^2 + \mathrm{o}((x-a)^2).$$

We will also write

$$f(x) \approx f(a) + f'(a)(x-a) + \frac{1}{2} f''(a)(x-a)^2$$

to indicate that (close to a) equality "almost" holds.

Sometimes in the formula for the Taylor polynomial we will put $x = a + h$, so we have

$$f(a+h) = f(a) + f'(a)h + \frac{1}{2} f''(a)h^2 + \mathrm{o}(h^2).$$

For example, for the function $f(t) = t \ln t$ (for which $f'(t) = \ln t + t\frac{1}{t} = \ln t + 1$, and $f''(t) = \frac{1}{t}$), we obtain

$$\begin{aligned}
(a+h)\ln(a+h) &= a \ln a + (\ln a + 1)h + \frac{1}{2}\frac{1}{a}h^2 + \mathrm{o}(h^2) \\
&\approx a \ln a + h \ln a + h + \frac{h^2}{2a}
\end{aligned}$$

which we will use below.

(b) Suppose the function f has a critical point at a: $f'(a) = 0$. Then

$$f(x) \approx f(a) + \frac{1}{2} f''(a)(x-a)^2$$

$$f(x) - f(a) \approx \frac{1}{2} f''(a)(x-a)^2.$$

Then (since a is a critical point) the difference $f(x) - f(a)$ has the same sign as $f''(a)$: if $f''(a) > 0$, f has a minimum at a; and if $f''(a) < 0$, in a there will be a maximum. If $f''(a) = 0$, we can say nothing.

Curvature

We have said above that the tangent line has the same *slope* as the graph of f, and that the best-fitting parabola has the same *curvature* as f. We now address the problem of the precise definition of curvature. We want to assign to each point on a curve, a number—the curvature at that point—which measures how fast the curve deviates from a straight line.

The first thing we can say about curvature is that it is *not* captured by the second derivative. Indeed, consider the parabola defined by $ax^2 + bx + c$:

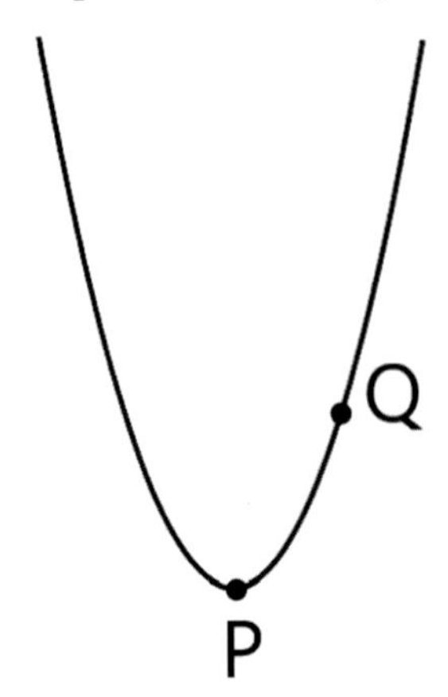

At all of its points its second derivative is the same: $2a$. Yet surely we would agree that its curvature—whatever it may be—is not the same at point P as at point Q.

So what exactly is curvature? Let's begin by considering circles:

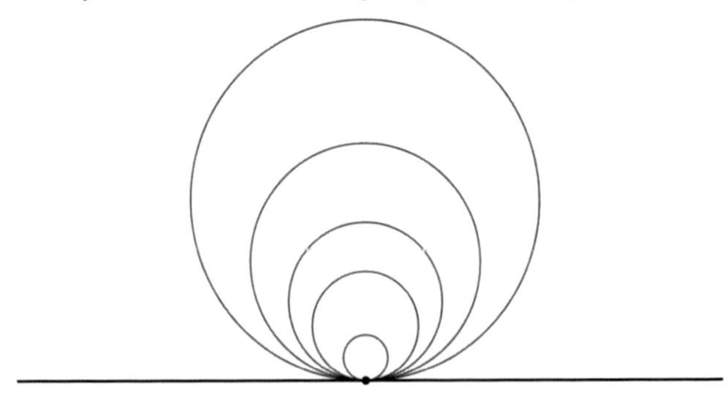

Two things we can agree on. The first is that a circle has the same curvature at all of its points: we can rotate a circle around its center, taking any of its points to any other...each has an analogous position with respect to the circle as a whole. The second is that smaller circles are more curved than larger circles. Consequently, we will use the radius of a circle to define its curvature at all of its points. We do this in the following way: at any point of a circle of radius r, the curvature of the circle is

$$\kappa = \frac{1}{r}.$$

Now we will use this to extend our definition of curvature to parabolas, and finally, to the graph of an arbitrary function f.

The Curvatures of a Parabola

We will define and calculate the curvature of a parabola at different points, say at points P and Q as in our picture above.

Consider first the point P, where the parabola has its minimum. Our parabola is given by $f(x) = ax^2$, and the point P is $(0, 0)$.

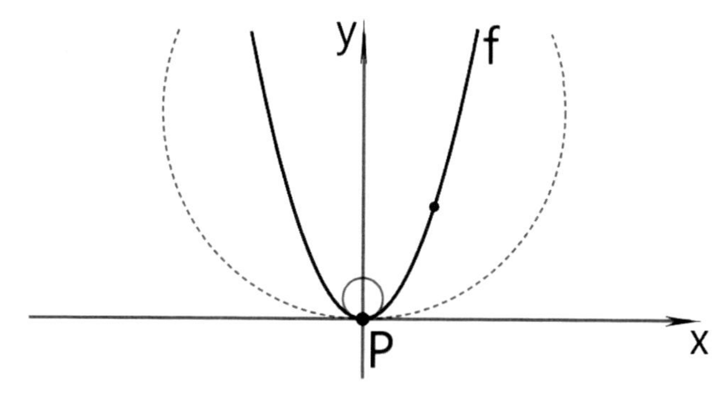

A small circle above the x-axis, centered on the y-axis and tangent at P will lie above the graph of f. But if we enlarge the circle, at some point this will no longer be so. Close to P, part of the circle will be below the parabola. The largest circle which remains above f close to the point P is called the *osculating circle* (from the Latin *osculum*=kiss) and we will consider its curvature as the curvature of f at P. Now let's calculate it. The circles with (large or small) radii r which we are interested in are those centered at $(0, r)$:

$$x^2 + (y - r)^2 = r^2.$$

And we seek the largest r such that $y \geq ax^2$ near $(0, 0)$. Thus we want the largest r for which

$$y = -\sqrt{r^2 - x^2} + r \geq ax^2.$$

Set $h(x) = -ax^2 - \sqrt{r^2 - x^2} + r$. This function h measures the vertical difference between the circle and the graph of f. Clearly $h(0) = 0$. For small values of r our discussion above says that this value must be a minimum, as the circle is above f. Differentiating h we have

$$h'(x) = -2ax + \frac{x}{\sqrt{r^2 - x^2}} = x\left(\frac{1}{\sqrt{r^2 - x^2}} - 2a\right)$$

so $h'(0) = 0$ and

$$h''(x) = \left(\frac{1}{\sqrt{r^2 - x^2}} - 2a\right) + x\left(\frac{1}{\sqrt{r^2 - x^2}} - 2a\right)'.$$

Thus

$$h''(0) = \frac{1}{r} - 2a$$

which will be positive for small r (and $h(0)$ will be a minimum) and negative for large r (and $h(0)$ will be a maximum: the circle is now below the parabola). The limiting case ($h''(0) = 0$) occurs when $r = \frac{1}{2a}$. Thus we conclude that the curvature of f at $P = (0, 0)$ is

$$\kappa = \frac{1}{r} = 2a = f''(0).$$

Now consider point Q of the parabola. To this end we take the parabola to be the graph of $f(x) = ax^2 + bx$, and $Q = (0, 0)$:

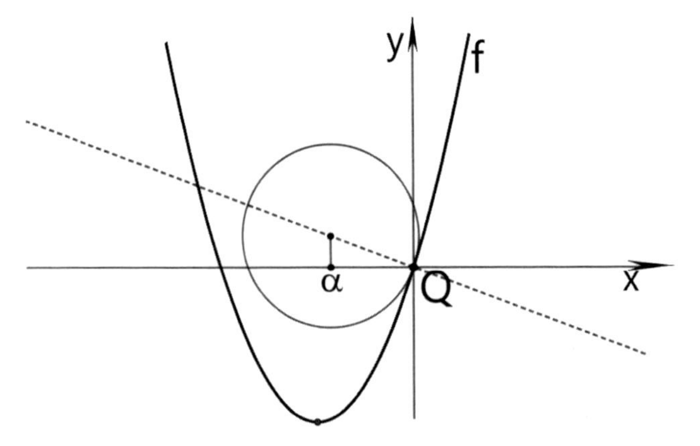

Our circles will now be centered on the line perpendicular to f at Q: $y = -\frac{1}{b}x$, (since $f'(0) = b$). The circle thus centered and of radius r is (note that α depends on r)

$$(x - \alpha)^2 + \left(y + \frac{\alpha}{b}\right)^2 = r^2 = \alpha^2 + \frac{\alpha^2}{b^2} = \alpha^2\left(1 + \frac{1}{b^2}\right).$$

We repeat our discussion above, and now we have

$$y = -\sqrt{\alpha^2\left(1+\frac{1}{b^2}\right) - (x-\alpha)^2} - \frac{\alpha}{b} \geq ax^2 + bx.$$

We set

$$h(x) = -ax^2 - bx - \sqrt{\alpha^2\left(1+\frac{1}{b^2}\right) - (x-\alpha)^2} - \frac{\alpha}{b}.$$

And we now seek the largest r such that h has a minimum at 0. Doing the computations (see Exercise 2) we now conclude that the limiting case ($h''(0) = 0$) occurs when

$$h''(0) = -2a - \frac{b(1+b^2)}{\alpha} = 0.$$

That is,

$$\alpha = -\frac{b(1+b^2)}{2a}.$$

But since $r^2 = \alpha^2(1 + \frac{1}{b^2})$,

$$r = |\alpha|\sqrt{1+\frac{1}{b^2}} = \frac{b(1+b^2)}{2a}\sqrt{\frac{1+b^2}{b^2}} = \frac{(1+b^2)^{3/2}}{2a}$$

and the curvature at point Q is then

$$\kappa = \frac{1}{r} = \frac{2a}{(1+b^2)^{3/2}},$$

which for the parabola f is

$$\kappa = \frac{f''(0)}{(1+f'(0)^2)^{3/2}}.$$

Note that the same formula holds for the point P, where we had $f'(0) = 0$. Accordingly, we define the curvature of the graph of f at the point $(x, f(x))$ as the curvature of the best-fitting parabola at that point

$$\kappa(x) = \frac{|f''(x)|}{(1+f'(x)^2)^{3/2}}.$$

The *signed* curvature is defined as $\frac{f''(x)}{(1+f'(x)^2)^{3/2}}$.

Random Walk and the Gauss Curve

Back in Chap. 1 we flipped a coin infinitely many times to generate a random point on the interval (0, 1) and to see that it was almost never rational. We will now use that coin to generate a "random walk": you stand at zero on the real line and toss a coin; if heads you take a step to your right, if tails, a step to your left. And now, from where you are, you do it again. And again. And again. And again... n times. You have had a random walk. Where have you arrived? Let's see. If you have thrown k heads, and therefore $n-k$ tails, you are at the point

$$k-(n-k)=-n+2k$$

(k steps to the right and $n-k$ steps to the left). You are at some whole number between $-n$ and n. But if, say, a is one of those points, What is the probability that you're standing at a? The total number of possible random walks is the number of possible results of throwing a coin n times: $2\times 2\times\cdots\times 2=2^n$. How many of these walks take you to the point $a=-n+2k$? To have reached that point you must have obtained $k=\frac{n+a}{2}$ heads in the n tosses. In other words, of the 2^n walks, $\binom{n}{k}$ take you to a. The probability of being there after n throws is therefore

$$\frac{\binom{n}{k}}{2^n}=\frac{\binom{n}{\frac{n+a}{2}}}{2^n}.$$

Note that if n is even, so is a, and if n is odd, so is a. Therefore $n+a$ will always be even.

If you have visited a science museum you have probably seen a "Galton board": a small ball falls, hits a nail and goes either right or left, hits another nail and again goes right or left, then another... n times. And the balls accumulate at the foot of the board forming a beautiful *Gauss curve*,

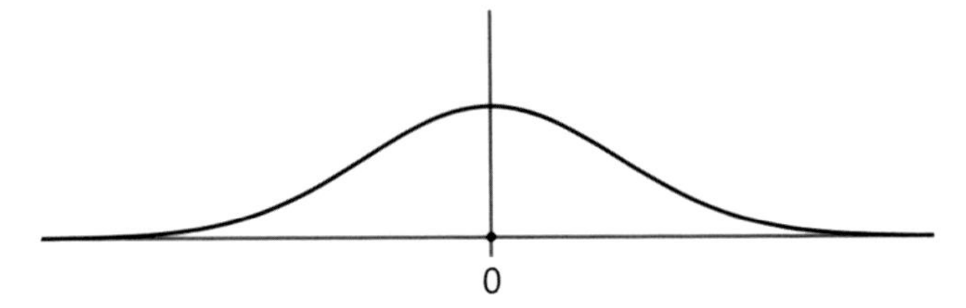

which is the graph of the function $f(x)=e^{-\frac{x^2}{2}}$. Magic! No: random walk. Let's see why. Say that 2^n balls are dropped (and each one of them has its random walk). According to our discussion above, the column of balls over point a should have a height of about

$$\binom{n}{\frac{n+a}{2}}$$

balls. To fix a unit of measure, we compare this column over point a with the column over 0 (the central column) which will have around $\binom{n}{\frac{n}{2}}$ balls. In this unit of measure, the height of the column over a is

$$H_a = \frac{\binom{n}{\frac{n+a}{2}}}{\binom{n}{\frac{n}{2}}} = \frac{\frac{n!}{\left(\frac{n+a}{2}\right)!\left(\frac{n-a}{2}\right)!}}{\frac{n!}{\left(\frac{n}{2}\right)!\left(\frac{n}{2}\right)!}} = \frac{\left(\frac{n}{2}\right)!\left(\frac{n}{2}\right)!}{\left(\frac{n+a}{2}\right)!\left(\frac{n-a}{2}\right)!}.$$

To facilitate our computations we will write $p = \frac{n}{2}$ and $c = \frac{a}{2}$. Thus, the height over a is

$$H_a = \frac{p!p!}{(p+c)!(p-c)!} = \frac{(p-c+1)\dots p}{(p+1)\dots(p+c)}$$

Taking ln,

$$\begin{aligned}
\ln H_a =& \sum_{j=p-c+1}^{p} \ln j - \sum_{j=p+1}^{p+c} \ln j \\
\approx& \int_{p-c}^{p} \ln t\, dt - \int_{p}^{p+c} \ln t\, dt \\
=&(t\ln t - t)\Big|_{p-c}^{p} - (t\ln t - t)\Big|_{p}^{p+c} \\
=&p\ln p - p - (p-c)\ln(p-c) + (p-c) - (p+c)\ln(p+c) + (p+c) + p\ln p - p \\
=&2p\ln p - (p-c)\ln(p-c) - (p+c)\ln(p+c).
\end{aligned}$$

Now we use the Taylor polynomial of order 2 of $t\ln t$ at p, which tells us that

$$(p+h)\ln(p+h) \approx p\ln p + h\ln p + h + \frac{h^2}{2p}:$$

$$\begin{aligned}
\ln H_a \approx& 2p\ln p - \left(p\ln p - c\ln p - c + \frac{c^2}{2p}\right) - \left(p\ln p + c\ln p + c + \frac{c^2}{2p}\right) \\
=& -\frac{c^2}{p}, \text{ which in terms of } n \text{ and } a \text{ is} \\
=& -\frac{a^2}{2n} \\
=& -\frac{\left[\frac{a}{\sqrt{n}}\right]^2}{2}
\end{aligned}$$

Then $H_a \approx e^{-\frac{\left[\frac{a}{\sqrt{n}}\right]^2}{2}}$. But, just us we "normalized" the height over 0 dividing by the number of balls, let's normalize the width of the board by setting $x = \frac{a}{\sqrt{n}}$ (in other words dividing the width which was $2n$ by $2\sqrt{n}$). When we now look at the board we see

$$e^{-\frac{x^2}{2}}.$$

Nothing more predictable than randomness, if it's repeated sufficiently many times. Hence, the importance of Statistics. The Gauss curve, or bell curve, appears in many real world situations which start with random walks: particles that move a little to one side and then to another; for example: the heat equation, and finance (where stocks and bonds go up a little, down a little...). We will see more on the Gauss curve and Normal distribution in Chap. 8.

Carl Friedrich Gauss (1777–1855) is generally considered one of the three greatest mathematicians (with Newton and Archimedes). He contributed an enormous number of results in many areas of mathematics: differential geometry, statistics, algebra; and also in mechanics, electrostatics, and geodesics. Among his results is the Fundamental Theorem of Algebra (every polynomial has a complex root), and his Theorema Egregium *on the curvature of surfaces.*

Before going on to Taylor series expansions, we give one more application of the Taylor polynomial of order two: when we saw the Newton–Raphson method for finding a zero, a, of the function f, we constructed a sequence (x_n). We may use the Taylor polynomial of order two of f to obtain a condition that will assure the convergence of that sequence (x_n) to the point a. Consider the Taylor polynomial of order two of f at x_n:

$$f(x) \approx f(x_n) + f'(x_n)(x - x_n) + \frac{f''(x_n)}{2}(x - x_n)^2.$$

At the point $x = a$, which is a zero of f,

$$0 = f(a) \approx f(x_n) + f'(x_n)(a - x_n) + \frac{f''(x_n)}{2}(a - x_n)^2.$$

If we divide by $f'(x_n)$, (recall that $x_{n+1} = x_n - \frac{f(x_n)}{f'(x_n)}$)

$$\begin{aligned} 0 &\approx \frac{f(x_n)}{f'(x_n)} + a - x_n + \frac{f''(x_n)}{2f'(x_n)}(a - x_n)^2 \\ &= (a - x_{n+1}) + \frac{f''(x_n)}{2f'(x_n)}(a - x_n)^2, \end{aligned}$$

from which

$$|x_{n+1} - a| \approx \frac{|f''(x_n)|}{2|f'(x_n)|}|x_n - a|^2.$$

Thus if close to point a we have that $\frac{|f''(x)|}{2|f'(x)|} \leq M$, and $|x_n - a|$ is small, then

$$|x_{n+1} - a| \leq M|x_n - a|^2$$

and the sequence (x_n) will converge to a.

The Taylor Series

Near the point $(a, f(a))$ the tangent line is the line that best approximates the function f, and the Taylor polynomial of order two is the parabola which best approximates f. Which would be the n-degree polynomial best approximating f? This will be the *Taylor polynomial of order n of f at a*. We had asked two things of the tangent line:

$$y(a) = f(a)$$
$$y'(a) = f'(a),$$

and three things of the Taylor polynomial of order two:

$$y(a) = f(a)$$
$$y'(a) = f'(a)$$
$$y''(a) = f''(a).$$

For the Taylor polynomial of order n we will require the following $n + 1$ conditions:

$$y(a) = f(a)$$
$$y'(a) = f'(a)$$
$$\vdots$$
$$y^{(n)}(a) = f^{(n)}(a)$$

in other words, that the derivatives of all orders up to n coincide with those of f at point a. Close to a, the difference between f and its Taylor polynomial of order n will be so small that (applying L'Hôpital n times..)

$$\frac{f(x) - y(x)}{(x-a)^n} \to \frac{0}{0}$$
$$\frac{f'(x) - y'(x)}{n(x-a)^{n-1}} \to \frac{0}{0}$$
$$\vdots$$
$$\frac{f^{(n)}(x) - y^{(n)}(x)}{n!} \to \frac{0}{n!} = 0,$$

so that $f(x) - y(x)$ tends to zero even when divided by $(x-a)^n$. We will write

$$f(x) = y(x) + \circ((x-a)^n).$$

But, how do we find this polynomial? Let's write

$$y(x) = c_0 + c_1(x-a) + c_2(x-a)^2 + \cdots + c_n(x-a)^n$$

and impose the conditions $y(a) = f(a)$, $y'(a) = f'(a), \ldots, y^{(n)}(a) = f^{(n)}(a)$. We will then obtain the necessary coefficients c_k. For example, after differentiating $y(x)$ three times:

$$y^{(3)}(x) = 3 \cdot 2 \cdot c_3 + 4 \cdot 3 \cdot 2 \cdot c_2(x-a) + \cdots + n(n-1)(n-2)c_n(x-a)^{n-3}$$
$$y^{(3)}(a) = 3!c_3$$

and we will have $c_3 = \frac{y^{(3)}(a)}{3!} = \frac{f^{(3)}(a)}{3!}$. Following in this manner we have, in general, that

$$c_k = \frac{f^{(k)}(a)}{k!}$$

so the Taylor polynomial of order n of f at a is

$$y(x) = \sum_{k=0}^{n} \frac{f^{(k)}(a)}{k!}(x-a)^k$$

and we will have

$$f(x) = \sum_{k=0}^{n} \frac{f^{(k)}(a)}{k!}(x-a)^k + \circ((x-a)^n).$$

Now, when you approximate something, it is very important to know the size of your error. Thus, we want to have some kind of bound for the difference $f(x) - y(x)$

between f and its Taylor polynomials. There are several formulas for this. Let's see one of them:

Say that b is a point close to a. We may write the difference at b as

$$f(b) - y(b) = M(b-a)^{n+1}.$$

Indeed, for this we need only put

$$M = \frac{f(b) - y(b)}{(b-a)^{n+1}}.$$

If we consider the function

$$g(x) = f(x) - y(x) - M(x-a)^{n+1},$$

we see that $g(a) = 0$. But also $g'(a) = 0$, $g''(a) = 0, \ldots, g^{(n)}(a) = 0$. This is because the derivatives of f and of y coincide at a, up to order n; and also the derivatives of $(x-a)^{n+1}$, up to order n, vanish at a. Now, from our definition of M, $g(b) = 0 = g(a)$, and by Rolle's Theorem, $g'(c_1) = 0$ for some c_1 between a and b. Thus, applying Rolle's Theorem repeatedly, $g'(c_1) = 0 = g'(a)$, then $g''(c_2) = 0$ with c_2 between a and c_1, $g''(c_2) = 0 = g''(a)$, then $g'''(c_3) = 0$ with c_3 between a and c_2, etc. ... until $g^{(n)}(c_n) = 0 = g^{(n)}(a)$, and then $g^{(n+1)}(c) = 0$ with c between a and c_n. But $g^{(n+1)}(x) = f^{(n+1)}(x) - M(n+1)!$, for y was an n-degree polynomial. Then we can write another expression for M:

$$0 = g^{(n+1)}(c) = f^{(n+1)}(c) - M(n+1)!$$

$$M = \frac{f^{(n+1)}(c)}{(n+1)!} \quad \text{with } c \text{ between } a \text{ and } b.$$

We then have $f(b) = y(b) + \frac{f^{(n+1)}(c)}{(n+1)!}(b-a)^{n+1}$, with c between a and b. But we can do this for any b, so

$$f(x) = y(x) + \frac{f^{(n+1)}(c)}{(n+1)!}(x-a)^{n+1},$$

with c between a and x (and c depending on x). Finally,

$$f(x) = \sum_{k=0}^{n} \frac{f^{(k)}(a)}{k!}(x-a)^k + \frac{f^{(n+1)}(c)}{(n+1)!}(x-a)^{n+1},$$

with the last term—the error—similar to the others, except that the derivative of f is evaluated at c instead of a... And c is a point between a and x, which depends on x, but is unknown to us.

At this stage one must ask oneself: What happens if we take larger and larger n? Will we have

$$f(x) = \sum_{k=0}^{\infty} \frac{f^{(k)}(a)}{k!}(x-a)^k ?$$

The answer is that even when f is differentiable infinitely many times, this series may or may not converge; or even, may converge but not to $f(x)$ (among the exercises, you will find an example of this). However, for many useful and interesting functions, including e^x, $\cos x$, $\sin x$, $\ln x$, all polynomials … the answer is yes: one may write, close to the point a, the *Taylor series expansion*:

$$f(x) = \sum_{k=0}^{\infty} \frac{f^{(k)}(a)}{k!}(x-a)^k.$$

Let's call $R_n(x)$ the difference between $f(x)$ and its Taylor polynomial of order n. We have seen above that

$$\begin{aligned} R_n(x) &= f(x) - \sum_{k=0}^{n} \frac{f^{(k)}(a)}{k!}(x-a)^k \\ &= \frac{f^{(n+1)}(c)}{(n+1)!}(x-a)^{n+1}, \end{aligned}$$

where $c \in (a, x)$. Then, if for any $c \in (a-r, a+r)$ and all $n \in \mathbb{N}$, we have

$$|f^{(n+1)}(c)| \leq K_r,$$

then we will have, for all $x \in (a-r, a+r)$

$$|R_n(x)| \leq \frac{K_r|x-a|^{n+1}}{(n+1)!} \leq K_r \frac{r^{n+1}}{(n+1)!}$$

and then $R_n(x) \longrightarrow 0$ for n tending to infinity, for $K_r \frac{r^{n+1}}{(n+1)!} \longrightarrow 0$. Thus, the Taylor series of f at a will converge to f in $(a-r, a+r)$.

Example Taylor series of $\frac{1}{1-x}$ at $a = 0$.

The derivatives of f:

$$f(x) = (1-x)^{-1} \qquad f(0) = 1$$

$$f'(x) = -(1-x)^{-2} \qquad f'(0) = 1$$

$$f''(x) = 2(1-x)^{-3} \qquad f''(0) = 2$$

$$f'''(x) = 3 \cdot 2(1-x)^{-4} \qquad f'''(0) = 3!$$

$$\vdots \qquad \vdots$$

$$f^{(k)}(x) = k!(1-x)^{-(k+1)} \qquad f^{(k)}(0) = k!$$

$$\frac{1}{1-x} = f(x) = \sum_{k=0}^{\infty} \frac{f^{(k)}(0)}{k!} x^k = \sum_{k=0}^{\infty} x^k.$$

Note that in this case the Taylor series at 0 converges to f in $(-1, 1)$. It does not converge, however, for x far from $a = 0$. For example, if $x = 2$ we have

$$-1 = \frac{1}{1-2} \neq \sum_{k=0}^{\infty} 2^k = \infty.$$

Example The Taylor series of $f(x) = \ln x$ at $a = 1$.

Derivatives of f:

$$f(x) = \ln x \qquad f(1) = 0$$

$$f'(x) = \frac{1}{x} = x^{-1} \qquad f'(1) = 1$$

$$f''(x) = -x^{-2} \qquad f''(1) = -1$$

$$f'''(x) = 2x^{-3} \qquad f'''(1) = 2$$

$$\vdots \qquad \vdots$$

$$f^{(k)}(x) = (-1)^{k+1}(k-1)!x^{-k} \qquad f^{(k)}(1) = (-1)^{k+1}(k-1)!$$

The Taylor series is

$$\sum_{k=0}^{\infty} \frac{f^{(k)}(1)}{k!}(x-1)^k = \sum_{k=1}^{\infty} \frac{(-1)^{k+1}}{k}(x-1)^k$$

which, by Leibniz' criterion for alternating series, converges if $|x - 1| < 1$. Let's consider the error term of order n at $x = 2$:

$$\begin{aligned}R_n(2) &= \frac{f^{(n+1)}(c)}{(n+1)!}(2-1)^{n+1}\\ &= \frac{(-1)^{n+2}n!}{(n+1)!c^{n+1}}\\ &= \frac{(-1)^n}{n+1}\frac{1}{c^{n+1}} \longrightarrow 0, \qquad \text{for } n \longrightarrow \infty \text{ and any } c \in (1,2).\end{aligned}$$

Thus, the Taylor series converges to the function on $x = 2$, and we have

$$\ln 2 = \sum_{k=1}^{\infty} \frac{(-1)^{k+1}}{k} = 1 - \frac{1}{2} + \frac{1}{3} - \frac{1}{4} + \frac{1}{5} - \dots$$

In Chap. 2, we saw that the alternating harmonic series was convergent, by Leibniz' criterion. Now we know it converges to $-\ln 2$.

The convergence of the Taylor series of a function f to the function f at all points may also happen. Functions for which this happens are called *entire*, and these include e^x, $\cos x$, $\sin x$. In a sense, these functions are surprisingly rigid: consider the graph of an entire function f:

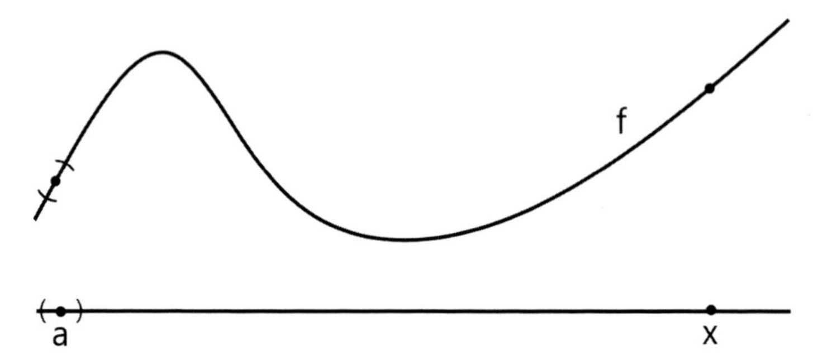

And consider that $f(x) = \sum_{k=0}^{\infty} \frac{f^{(k)}(a)}{k!}(x-a)^k$. On the right-hand side of the equality, all the "information" about the function f is contained in its derivatives $f^{(k)}(a)$ at the point a. And these depend only on what happens in a neighborhood of a, as small as we want ... This means that if we know f in any interval around a, we know f everywhere.

Example The Taylor series of $f(x) = \sin x$ at $a = 0$.

Derivatives of f:

$$\begin{aligned} f(x) &= \sin x & f(0) &= 0\\ f'(x) &= \cos x & f'(0) &= 1\\ f''(x) &= -\sin x & f''(0) &= 0 \end{aligned}$$

$$f'''(x) = -\cos x \qquad f'''(0) = -1$$

$$f^{(4)}(x) = \sin x \qquad f^{(4)}(0) = 0$$

$$\vdots \qquad \vdots$$

Note that for any value of c and for all $n \in \mathbb{N}$, we have

$$|f^{(n+1)}(c)| \leq 1.$$

Then,

$$|R_n(x)| \leq \frac{|x|^{n+1}}{(n+1)!} \longrightarrow 0 \quad \text{for any } x \in \mathbb{R}.$$

Thus, the Taylor series of $\sin x$ at 0 tends to $\sin x$ at all points:

$$\sin x = \sum_{k=0}^{\infty} \frac{(-1)^k}{(2k+1)!} x^{2k+1} = x - \frac{1}{3!}x^3 + \frac{1}{5!}x^5 - \frac{1}{7!}x^7 + \ldots$$

Exercises

1 Prove that a is a double root of a polynomial P if and only if a is a root of P and of P'.

2 Complete the computations of the discussion of curvature of the parabola, by calculating the second derivative of

$$h(x) = -ax^2 - bx - \sqrt{\alpha^2\left(1 + \frac{1}{b^2}\right) - (x - \alpha)^2} - \frac{\alpha}{b}.$$

3 Calculate the curvature of

(i) $f(x) = x^2$ at $(0, 0)$.
(ii) $f(x) = x^4$ at $(0, 0)$. Interpret...
(iii) a circle of radius r at any of its points, by applying the curvature formula to $f(x) = \sqrt{r^2 - x^2}$.

4 Curvature of a parametrized curve: Let $p(t) = (x(t), y(t))$, where $y(t) = f(x(t))$. By using the chain rule translate the curvature formula to obtain

$$\kappa = \frac{|x'y'' - x''y'|}{(x'^2 + y'^2)^{3/2}}.$$

5 Calculate the Taylor polynomial of order three of $\frac{1}{x+1}$ at $a = 0$. Setting $x = \frac{1}{10}$, estimate the decimal expression of $\frac{10}{11}$.

6 Calculate the Taylor series at $a = 0$ of the following functions, and study their convergence.

i) $\cos x$.
ii) $\sqrt{x+1}$.
iii) $\frac{1}{1+x^2}$.

7 Prove that all derivatives of the function

$$f(x) = \begin{cases} e^{-\frac{1}{x^2}}, & \text{if } x \neq 0 \\ 0, & \text{if } x = 0 \end{cases}$$

vanish at zero. Conclude that the Taylor series of f at $a = 0$ converges, but not to f.

7 Convexity and the Isoperimetric Inequality

In this chapter we consider some important inequalities. We begin with the Arithmetic-Geometric inequality and convexity and finally present the Isoperimetric Inequality.

The Arithmetic-Geometric Inequality

The Arithmetic-Geometric inequality (or AG inequality) is one of the simplest and most useful inequalities, as well as the source of many other important variations. We have seen among the exercises of Chap. 4 that it can be used to solve some optimization problems without the use of differentiation. The AG inequality says that if a and b are positive numbers, then

$$\sqrt{ab} \leq \frac{a+b}{2},$$

that is, the *geometric mean* is always less than the *arithmetic mean*. The proof is simple: given any two numbers x and y,

$$\begin{aligned} 0 &\leq (x-y)^2 \\ &= x^2 - 2xy + y^2, \\ \text{from where} \quad 2xy &\leq x^2 + y^2 \\ xy &\leq \frac{x^2+y^2}{2}, \end{aligned}$$

thus, given positive a and b, we may take $x = \sqrt{a}$ and $y = \sqrt{b}$, to obtain $\sqrt{ab} \leq \frac{a+b}{2}$. Note that equality occurs only if a and b are equal.

I. Zalduendo, *Calculus off the Beaten Path*, SUMS Readings,
https://doi.org/10.1007/978-3-031-15765-3_7

Convexity

After the definition of the Taylor polynomial of order two of f, our second comment supposed that a was a critical point. Even without supposing this, we can write

$$f(x) \approx f(a) + f'(a)(x-a) + \frac{1}{2}f''(a)(x-a)^2, \text{ so}$$

$$f(x) - [f(a) + f'(a)(x-a)] \approx \frac{1}{2}f''(a)(x-a)^2$$

which shows that the relative position of the graph of f and of its tangent lines is linked to the sign of the second derivative of f. The notion of *convexity* of a function is very important in analysis and in many applications. It can be expressed in several ways:

Theorem *The following are equivalent:*

(i) the lines tangent to the graph of f remain under the graph,
(ii) for all $x < y$ and all $0 \leq t \leq 1$,

$$f(tx + (1-t)y) \leq tf(x) + (1-t)f(y),$$

(iii) f' is increasing,
(iv) $f'' \geq 0$.

We will show that (i) implies (ii), (ii) implies (iii), (iii) implies (i). And clearly, we already know that (iii) is equivalent to (iv).

That (i) implies (ii): Take $a = tx + (1-t)y$, and note that then

$$x - a = x - tx - (1-t)y = (1-t)(x-y)$$

$$y - a = y - tx - (1-t)y = t(y-x).$$

As all tangent lines remain below f,

$$f(a) + f'(a)(x-a) = f(a) + f'(a)(1-t)(x-y) \leq f(x)$$

$$f(a) + f'(a)(y-a) = f(a) + f'(a)t(y-x) \leq f(y).$$

Now, if we multiply the first inequality by t, the second by $(1-t)$ and add:

$$f(tx + (1-t)y) = f(a) \leq tf(x) + (1-t)f(y).$$

That (ii) implies (iii): Say that $w < z$, and take x and y such that

$$w < x < y < z.$$

We first consider the three points to the right ($x < y < z$), and write $y = tx + (1-t)z$; we can do this by setting $t = \frac{z-y}{z-x}$, and $1 - t = \frac{y-x}{z-x}$. Then by ii)

$$\begin{aligned} f(y) = f(tx + (1-t)z) &\leq tf(x) + (1-t)f(z) \\ &= \left(\frac{z-y}{z-x}\right) f(x) + \left(\frac{y-x}{z-x}\right) f(z), \end{aligned}$$

from where

$$f(y)(z-x) \leq f(x)(z-y) + f(z)(y-x) \tag{$*$}$$

Now, adding and subtracting x in $(z-y)$:

$$\begin{aligned} &= f(x)((z-x) + (x-y)) + f(z)(y-x) \\ &= f(x)(z-x) + f(x)(x-y) + f(z)(y-x). \end{aligned}$$

Then

$$\begin{aligned} (f(y) - f(x))(z-x) &\leq (f(z) - f(x))(y-x), \text{ and} \\ \frac{f(y) - f(x)}{y-x} &\leq \frac{f(z) - f(x)}{z-x}. \end{aligned} \tag{7.1}$$

If, instead, in $(*)$ we add and subtract z in $(y-x)$, we analogously obtain

$$\frac{f(z) - f(x)}{z-x} \leq \frac{f(z) - f(y)}{z-y}. \tag{7.2}$$

And now, from (7.1) and (7.2):

$$\frac{f(y) - f(x)}{y-x} \leq \frac{f(z) - f(y)}{z-y}. \tag{A}$$

We now consider the three points to the left ($w < x < y$) and do exactly the same, to obtain

$$\frac{f(x) - f(w)}{x-w} \leq \frac{f(y) - f(x)}{y-x}. \tag{B}$$

From (A) and (B):

$$\frac{f(x) - f(w)}{x-w} \leq \frac{f(z) - f(y)}{z-y} = \frac{f(y) - f(z)}{y-z}.$$

Now, having $x \to w$ and $y \to z$,

$$f'(w) \leq f'(z),$$

so f' is increasing.

Finally that iii) implies i): Measure the difference $g(x)$ between the graph of f and the tangent line at the point $(a, f(a))$:

$$\begin{aligned} g(x) &= f(x) - f(a) - f'(a)(x-a) \\ &= f'(c)(x-a) - f'(a)(x-a) \\ &= (f'(c) - f'(a))(x-a) \end{aligned}$$

where in the second equality we used Lagrange's Mean Value Theorem. Note that c is between a and x. If $x > a$ (and therefore $c > a$ and $f'(c) \geq f'(a)$), $g(x)$ is the product of two positive numbers: $g(x) \geq 0$. If $x < a$ (and therefore $c < a$ and $f'(c) \leq f'(a)$), $g(x)$ is the product of two negative numbers: $g(x) \geq 0$. In any case, the tangent line remains below the graph of f. This completes the proof. □

Some comments:

(a) A function is said to be *convex* if it verifies the equivalent conditions in the Theorem, and *concave* if it verifies the "opposite" conditions: tangent lines below the graph, $\geq$ in ii), f' decreasing, $f'' \leq 0$. Note that with condition ii) we can define convexity even for non-differentiable functions.

(b) We may think of convexity of f as the following property: "f applied to an arithmetic mean of x and y is less than an arithmetic mean of $f(x)$ and $f(y)$." There is a related notion, but stronger than convexity, called *log-convexity*. A function f is said to be *log-convex* if the function $\ln f(x)$ is convex. Let's see what this means: it must happen, for $x < y$ and $0 \leq t \leq 1$ that

$$\ln(f(tx + (1-t)y)) \leq t \ln f(x) + (1-t) \ln f(y).$$

If we apply the exponential function to this inequality, we find that this is equivalent to

$$f(tx + (1-t)y) \leq f(x)^t f(y)^{(1-t)},$$

which we may think of as: "f applied to an arithmetic mean of x and y is less than a geometric mean of $f(x)$ and $f(y)$." Since geometric means are smaller than arithmetic means, it turns out that if f is log-convex, then it is also convex:

$$f(tx + (1-t)y) \leq f(x)^t f(y)^{(1-t)} \leq tf(x) + (1-t)f(y).$$

Thus, log-convexity is a strong form of convexity.

As applications, we mention some other important inequalities.

Young, Hölder, Jensen, Cauchy–Schwarz...

We start with the inequalities of Young and Heinz. These are generalizations of the Arithmetic-Geometric inequality.

Young's Inequality *if* $\frac{1}{p} + \frac{1}{q} = 1$, $ab \le \frac{a^p}{p} + \frac{b^q}{q}$.
Let's see

$$\begin{aligned} ab &= \mathrm{e}^{\ln(ab)} \\ &= \mathrm{e}^{\ln a + \ln b} \\ &= \mathrm{e}^{\frac{1}{p} p \ln a + \frac{1}{q} q \ln b} \\ &= \mathrm{e}^{\frac{1}{p} \ln a^p + \frac{1}{q} \ln b^q}. \end{aligned}$$

But e^x is a convex function, for $(\mathrm{e}^x)'' = \mathrm{e}^x > 0$. Then, since $\frac{1}{p} + \frac{1}{q} = 1$,

$$\begin{aligned} &\le \frac{1}{p}\mathrm{e}^{\ln(a^p)} + \frac{1}{q}\mathrm{e}^{\ln(b^q)} \\ &= \frac{a^p}{p} + \frac{b^q}{q}. \end{aligned}$$

□

Another way (which we have mentioned above) of writing Young's inequality is

$$a^t b^{1-t} \le ta + (1-t)b \quad \text{si } 0 \le t \le 1.$$

To see this, take $p = \frac{1}{t}$ and $q = \frac{1}{1-t}$. Set $x = a^{\frac{1}{p}}$ and $y = b^{\frac{1}{q}}$. Then

$$a^t b^{1-t} = xy \le \frac{x^p}{p} + \frac{y^q}{q} = ta + (1-t)b.$$

Heinz' Inequality

$$\sqrt{ab} \le \frac{a^t b^{1-t} + a^{1-t} b^t}{2} \le \frac{a+b}{2}.$$

Because

$$\sqrt{ab} = \sqrt{a^t a^{1-t} b^t b^{1-t}} = \sqrt{\left(a^t b^{1-t}\right)\left(a^{1-t} b^t\right)}$$

$$\leq \frac{a^t b^{1-t} + a^{1-t} b^t}{2}$$
$$\leq \frac{ta + (1-t)b + (1-t)a + tb}{2}$$
$$= \frac{a+b}{2}$$

the first inequality is AG, the second is Young's. □

The Arithmetic-Geometric Inequality for n Numbers

$$\sqrt[n]{a_1 \cdots a_n} \leq \frac{a_1 + \cdots + a_n}{n}.$$

Note that $\ln x$ is a concave function. Indeed, if we differentiate it twice,

$$(\ln x)'' = \frac{-1}{x^2} < 0.$$

Now, if we consider positive numbers $a_1, a_2, \ldots, a_n$,

$$\ln\left(\frac{1}{n}\sum_{i=1}^{n} a_i\right) = \ln\left(\sum_{i=1}^{n} \frac{1}{n} a_i\right)$$
$$\geq \frac{1}{n}\sum_{i=1}^{n} \ln(a_i)$$
$$= \ln\left(\left[\prod_{i=1}^{n} a_i\right]^{\frac{1}{n}}\right).$$

Applying e^x,

$$\frac{a_1 + \cdots + a_n}{n} \geq \sqrt[n]{a_1 \cdots a_n}.$$

□

Hölder's Inequality If f is a continuous function defined on the interval $[a, b]$, we define for any value of r (with $1 \leq r < \infty$),

$$\|f\|_r = \left(\int_a^b |f(t)|^r \, dt\right)^{\frac{1}{r}}.$$

Note that for $r = 2$ this is a generalization to "many variables" of the way we measure distances on the plane: $d((x_1, x_2); (0, 0)) = \sqrt{x_1^2 + x_2^2}$. Doing this with different values of r, represents different ways of measuring the "size" of a function f. This is important in many applications which we will not go into here. What Hölder's inequality then says is the following:

$$\text{If } \frac{1}{p} + \frac{1}{q} = 1, \text{ then } \|fg\|_1 \leq \|f\|_p \|g\|_q.$$

To prove this, note first that dividing f by $\|f\|_p$, and g by $\|g\|_q$, we may suppose that $\|f\|_p = \|g\|_q = 1$. Now by Young's inequality, we have, for all $t \in [a, b]$,

$$|f(t)||g(t)| \leq \frac{|f(t)|^p}{p} + \frac{|g(t)|^q}{q}.$$

Integrating on both sides,

$$\begin{aligned}
\|fg\|_1 = \int_a^b |f(t)||g(t)|\, dt &\leq \int_a^b \frac{|f(t)|^p}{p}\, dt + \int_a^b \frac{|g(t)|^q}{q}\, dt \\
&= \frac{1}{p}\|f\|_p^p + \frac{1}{q}\|g\|_q^q \\
&= \frac{1}{p} + \frac{1}{q} \\
&= 1 \\
&= \|f\|_p \|g\|_q.
\end{aligned}$$

□

Two more inequalities valid for convex functions: Jensen's and the Hermite-Hadamard inequality:

Jensen's Inequality *If f is convex and $0 < a_i < 1$, Then*

$$f\left(\frac{\sum_{i=1}^n a_i x_i}{\sum_{i=1}^n a_i}\right) \leq \frac{\sum_{i=1}^n a_i f(x_i)}{\sum_{i=1}^n a_i}.$$

Let's see why. Suppose first that $\sum_{i=1}^n a_i = 1$. We proceed by induction on n. For $n = 2$ (note that $a_1 + a_2 = 1$ means that $a_2 = 1 - a_1$):

$$f(a_1 x_1 + a_2 x_2) \leq a_1 f(x_1) + a_2 f(x_2).$$

For $n > 2$, our inductive hypothesis tells us that the conclusion is valid for $n - 1$ terms. Note that $\sum_{i=1}^{n-1} a_i = 1 - a_n$ and write

$$\begin{aligned}
f\left(\sum_{i=1}^{n} a_i x_i\right) &= f\left(\sum_{i=1}^{n-1} a_i x_i + a_n x_n\right) \\
&= f\left((1-a_n)\frac{\sum_{i=1}^{n-1} a_i x_i}{(1-a_n)} + a_n x_n\right) \\
&\leq (1-a_n) f\left(\sum_{i=1}^{n-1} \frac{a_i}{1-a_n} x_i\right) + a_n f(x_n) \\
&\leq (1-a_n) \sum_{i=1}^{n-1} \frac{a_i}{1-a_n} f(x_i) + a_n f(x_n) \\
&= \sum_{i=1}^{n} a_i f(x_i).
\end{aligned}$$

Now, if $\sum_{i=1}^{n} a_i$ is not one, replace each a_j by $\frac{a_j}{\sum_{i=1}^{n} a_i}$. □

The Hermite-Hadamard Inequality *If f is convex,*

$$f\left(\frac{a+b}{2}\right) \leq \frac{1}{b-a}\int_a^b f \leq \frac{f(a)+f(b)}{2}.$$

To see the first inequality, call $p = \frac{a+b}{2}$. We know that the tangent line at $(p, f(p))$ remains below the graph of f:

$$\begin{aligned}
f(p) + f'(p)(x-p) &\leq f(x) \\
f(p) + f'(p)x - f'(p)p &\leq f(x).
\end{aligned}$$

Integrating between a and b:

$$\begin{aligned}
f(p)(b-a) + f'(p)\frac{x^2}{2}\Big|_a^b - f'(p)p(b-a) &\leq \int_a^b f(x)\,dx \\
f(p)(b-a) + f'(p)\frac{b^2-a^2}{2} - f'(p)\frac{(a+b)}{2}(b-a) &\leq \int_a^b f(x)\,dx \\
f(p)(b-a) + f'(p)\frac{1}{2}(b^2-a^2) - f'(p)\frac{1}{2}(a+b)(b-a) &\leq \int_a^b f(x)\,dx
\end{aligned}$$

$$f(p)(b-a) \leq \int_a^b f(x)\,dx$$
$$f\left(\frac{a+b}{2}\right) = f(p) \leq \frac{1}{b-a}\int_a^b f(x)\,dx.$$

To see the second inequality, use the change of variables

$$x : [0, 1] \to [a, b] \text{ such that } x(t) = tb + (1-t)a,$$

and bear in mind that $x'(t) = b - a$, so

$$\begin{aligned}
\frac{1}{b-a}\int_a^b f(x)\,dx &= \frac{1}{b-a}\int_0^1 f(tb + (1-t)a)(b-a)\,dt \\
&= \int_0^1 f(tb + (1-t)a)\,dt \\
&\leq \int_0^1 tf(b) + (1-t)f(a)\,dt \\
&= f(b)\int_0^1 tdt + f(a)\int_0^1 1 - t\,dt \\
&= f(b)\frac{t^2}{2}\Big|_0^1 + f(a) - f(a)\frac{t^2}{2}\Big|_0^1 \\
&= \frac{f(b)}{2} + f(a) - \frac{f(a)}{2} \\
&= \frac{f(a) + f(b)}{2}.
\end{aligned}$$

□

Lastly, we will need the following inequality when we prove the isoperimetric inequality in the next section.

The Cauchy–Schwarz Inequality *This is the inequality*

$$|xw + yz| \leq \sqrt{x^2 + y^2}\,\sqrt{w^2 + z^2}.$$

Let's prove it. Applying the AG inequality to $(xz)^2$ and $(yw)^2$ we obtain

$$xywz = \sqrt{(xz)^2(yw)^2} \leq \frac{x^2z^2 + y^2w^2}{2}$$
$$\text{which is} \quad 2xywz \leq x^2z^2 + y^2w^2.$$

Summing $x^2w^2 + y^2z^2$, we have

$$
\begin{aligned}
2xywz + x^2w^2 + y^2z^2 &\leq x^2z^2 + y^2w^2 + x^2w^2 + y^2z^2 \\
(xw + yz)^2 &\leq x^2\left(w^2 + z^2\right) + y^2\left(w^2 + z^2\right) \\
&= \left(x^2 + y^2\right)\left(w^2 + z^2\right),
\end{aligned}
$$

and taking square root:

$$|xw + yz| \leq \sqrt{x^2 + y^2}\sqrt{w^2 + z^2}.$$

Note that the equality holds if and only if $|xz| = |yw|$, in other words,

$$\frac{x}{y} = \pm\frac{w}{z}.$$

□

The Isoperimetric Inequality

This is the inequality

$$A \leq \frac{L^2}{4\pi},$$

where, given any plane figure, A is its area, and L its perimeter. The meaning of this important inequality is that of all plane figures with a given perimeter, the one with the largest area is the circle: observe that for a circle, the equality holds.

But let's start with the case of rectangles. The analogous property for rectangles is that of all rectangles with a given perimeter, the one with the largest area is the square. This property is equivalent to the AG inequality.

To see this equivalence, suppose first that the AG inequality holds, and consider a rectangle, say of base b and height a. Its area is ab. And the AG inequality tells us (when squared) that

$$\text{area of the rectangle} = ab \leq \left(\frac{a+b}{2}\right)^2 = \left(\frac{2(a+b)}{4}\right)^2,$$

which is the area of the square of side $\frac{2(a+b)}{4}$ (the one with the same perimeter as the rectangle).

Let's try the other way around. So now suppose that the isoperimetric property for rectangles holds: of all rectangles with a given perimeter, the one with largest area is the square. Take positive numbers a and b (suppose $a \leq b$). Consider now the rectangle

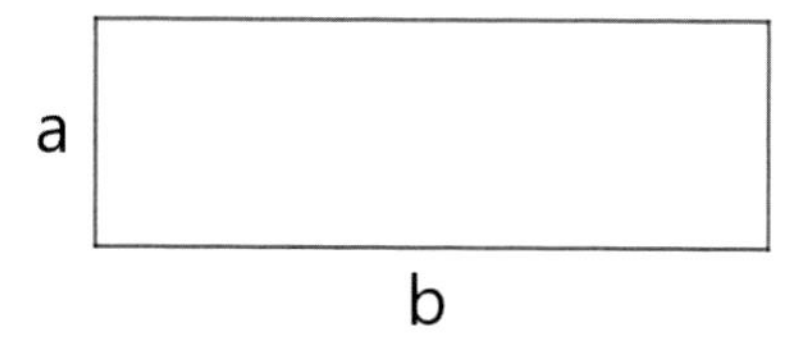

and "reorder" the lengths of its sides to obtain a square with the same perimeter:

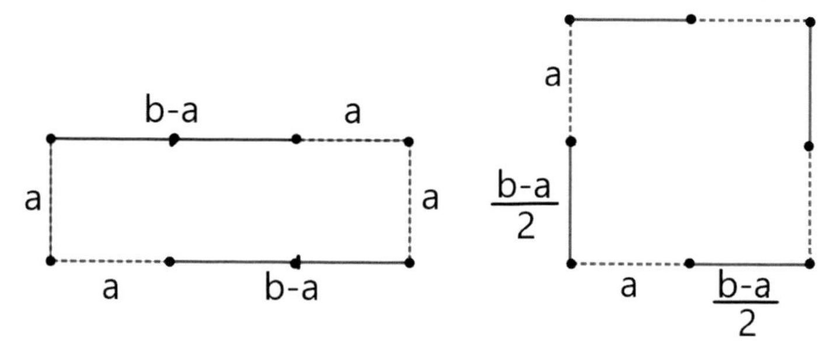

Compare their areas:

$$\text{area of the rectangle} \leq \text{area of the square}$$

$$\begin{aligned} ab &\leq \left(a + \frac{b-a}{2}\right)^2 \\ &= \left(\frac{2a+b-a}{2}\right)^2 \\ &= \left(\frac{a+b}{2}\right)^2, \end{aligned}$$

which, taking square roots, gives us the Arithmetic-Geometric inequality.

Note that the isoperimetric inequality for rectangles may be written as

$$A \leq \frac{L^2}{16}.$$

Since $4\pi < 16$, this is "finer" than the general isoperimetric inequality, but of course, it is only applicable to rectangles... Finer still, but only applicable to triangles, is the following.

Santaló's Inequality

$$A \leq \frac{L^2}{12\sqrt{3}},$$

which is also related to the AG inequality, in its three-number version: $\sqrt[3]{abc} \leq \frac{a+b+c}{3}$. To prove Santaló's inequality, write the area A of a triangle in terms of its semiperimeter s. The semiperimeter is half the sum of its sides: $s = \frac{L}{2} = \frac{\alpha+\beta+\gamma}{2}$. Heron's formula, from the elementary geometry that they should have taught us at school, says that the area of the triangle is

$$A = \sqrt{s(s-\alpha)(s-\beta)(s-\gamma)}.$$

Thus,

$$\begin{aligned} A^2 &= s(s-\alpha)(s-\beta)(s-\gamma) \\ &\leq s\left(\frac{s-\alpha+s-\beta+s-\gamma}{3}\right)^3 \\ &= s\left(\frac{3s-2s}{3}\right)^3 \\ &= s\left(\frac{s}{3}\right)^3 \\ &= \frac{s^4}{27}, \end{aligned}$$

from which, taking square roots,

$$A \leq \frac{s^2}{3\sqrt{3}} = \frac{L^2}{12\sqrt{3}},$$

which is Santaló's inequality. □

Note that equality holds when $\alpha = \beta = \gamma$, that is, of all triangles with a given perimeter, the one with the largest area is the equilateral triangle.

We now come to the general isoperimetric inequality. The proof we give is due to Erhard Schmidt [11], as it appears in Manfredo Do Carmo [7].

Isoperimetric Inequality *Given a plane region of area A and perimeter L,*

$$A \leq \frac{L^2}{4\pi}.$$

Let's prove this. Consider a region of area A, enclosed by a simple closed curve C of length L. We center the region on the y-axis, and draw a circumference centered at $(0, 0)$ of the same width (say $2r$) as our region.

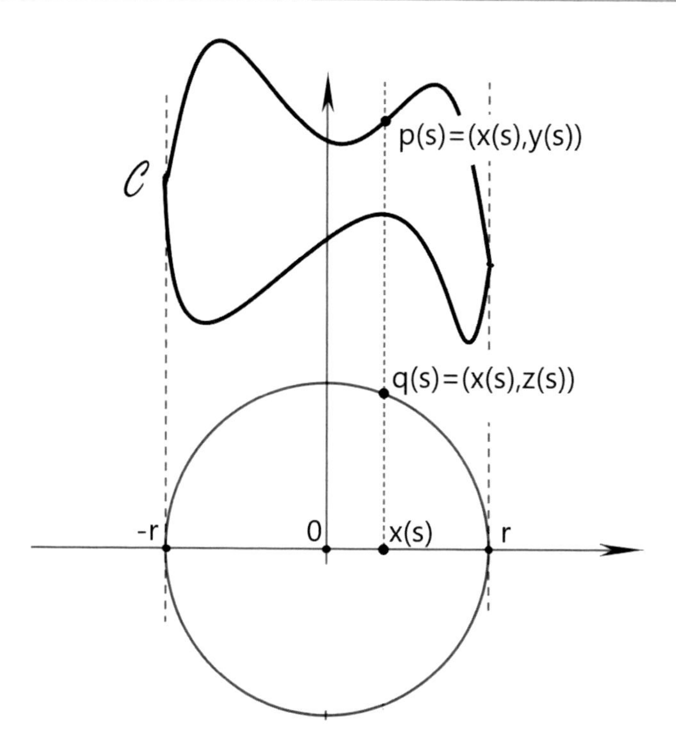

Parametrize $\mathcal{C}$ by arc length: $p : [0, L] \longrightarrow \mathbb{R}^2$ so that $p(s) = (x(s), y(s))$ (see Exercise 15 of Chap. 5). Now parametrize the circumference using the same first coordinate $x(s)$, that is: $q : [0, L] \longrightarrow \mathbb{R}^2$ with $q(s) = (x(s), z(s))$, where $z(s) = \pm\sqrt{r^2 - x(s)^2}$. Note that we then have: $x^2 + z^2 = r^2$, and $x'^2 + y'^2 = 1$. The formulas for the area enclosed by a curve (Chap. 5) tell us that

$$\pi r^2 = \int_0^L zx' \quad \text{and} \quad A = -\int_0^L xy'.$$

We then have, using the AG inequality,

$$\begin{aligned} 2\sqrt{\pi r^2}\sqrt{A} \leq \pi r^2 + A &= \int_0^L zx' - \int_0^L xy' \\ &= \int_0^L (zx' - xy') \\ &\leq \int_0^L \sqrt{z^2 + x^2}\sqrt{x'^2 + y'^2} \\ &= \int_0^L r \cdot 1 \end{aligned}$$

$$= rL,$$

where the last inequality is by Cauchy–Schwarz. Squaring, we have $4\pi r^2 A \leq r^2 L^2$, from which

$$A \leq \frac{L^2}{4\pi}.$$

□

Note that for equality to hold, we must have $\frac{x}{y'} = \pm\frac{z}{x'}$, that is,

$$xx' = \pm y'\sqrt{r^2 - x^2},$$

a differential equation verified by $(x(s), y(s)) = (r\cos(\frac{s}{r}), c + r\sin(\frac{s}{r}))$: a circumference of radius r centered at $(0, c)$.

Exercises

1 Prove that f is convex if and only if $-f$ is concave.

2 Prove that the sum of convex functions is convex.

3 Prove the inequality:

$$x \ln x + y \ln y \geq (x + y) \ln\left(\frac{x + y}{2}\right), \qquad \text{for all } x, y > 0.$$

4 Prove that if f and g are convex, then the function $h(x) = \max\{f(x), g(x)\}$ is convex. What about $\min\{f(x), g(x)\}$?

5 We have mentioned that even for non-differentiable functions, convexity may be defined by: ii) for all $x < y$ and all $0 \leq t \leq 1$,

$$f(tx + (1 - t)y) \leq tf(x) + (1 - t)f(y).$$

Prove that such a function will always be continuous (Hint: see ii) $\Rightarrow$ iii) of the equivalences).

6 Let $\alpha_1, \ldots, \alpha_n > 0$ be such that $\sum_{i=1}^{n} \alpha_i = 1$. Prove the generalized Arithmetic-Geometric inequality:

$$a_1^{\alpha_1} \cdots a_n^{\alpha_n} \leq \alpha_1 a_1 + \cdots + \alpha_n a_n, \qquad \text{for } a_i > 0.$$

7 Prove that

(i) x^a is log-convex if $a \leq 0$.
(ii) x^a is log-concave if $a \geq 0$.
(iii) e^{-x^2} is log-concave.

8 Prove that if f and g are log-concave (resp. log-convex) then their product, fg, is log-concave (resp. log-convex).

9 A subset A of the plane is said to be *convex* if given any two points $P, Q \in A$, the segment joining them, $\overline{PQ}$ is completely contained in A. Given a set A, whose boundary is a closed curve (parametrized anti-clockwise, as in the figure by $p(t) = (x(t), y(t))$),

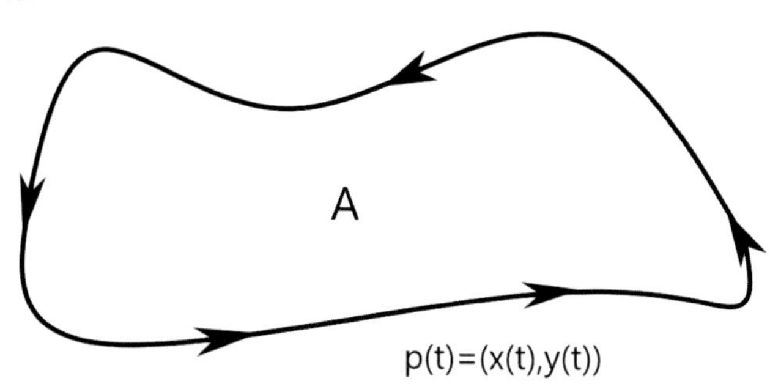

(i) Show that A is convex if and only if $x'y'' - y'x'' > 0$.
(ii) Deduce that the convexity of A can be determined knowing only its boundary.

10 Show that the sequence $x_n = \left(1 + \frac{1}{n}\right)^n$ is increasing. Hint: use the AG inequality for the $n+1$ numbers

$$1, \left(1 + \frac{1}{n}\right), \ldots, \left(1 + \frac{1}{n}\right)$$

More Integrals

8

We give here some applications of the integral to the calculation of volumes and surface areas. Density functions, expectation and barycenter are also briefly considered.

Volume

We have seen in Chap. 3 how Eudoxus and other Greek mathematicians calculated areas and volumes by the process of exhaustion, approximating from within and without the area which they wanted to calculate. In the XVIIth Century Bonavantura Cavalieri (1598–1647) had an idea that was strongly criticized at the time: he considered an area as a "sum of lines" and a volume as a "sum of areas." Thus, for example, the area under the parabola $y = x^2$ between 0 and a (which for us is $\int_0^a x^2\,dx$) was for Bonaventura

$$\text{omn.}x^2 \qquad \text{(omnes lineae = all the lines)},$$

but while we calculate this using Barrow's rule, Bonavantura compared with other "omnes lineae" already known to him. For example, if we consider the pyramid of square base $a \times a$, and height a, the area of a horizontal cut at distance x from the vertex (which is a square of sides x) is x^2. Thus the pyramid (whose volume is $\frac{a^3}{3}$), has volume = "sum of areas".

I. Zalduendo, *Calculus off the Beaten Path*, SUMS Readings,
https://doi.org/10.1007/978-3-031-15765-3_8

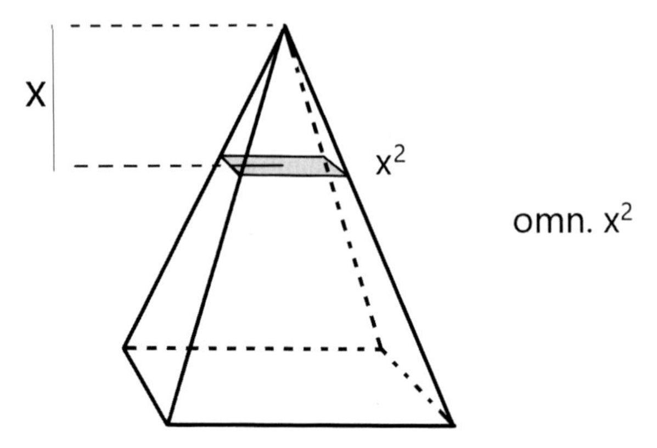

and therefore the area under the parabola $y = x^2$ between 0 and a is also $\frac{a^3}{3}$.

If in comparing two bodies A and B, Cavalieri found that the areas of the corresponding sections were proportional:

$$\mathrm{B}_x = k\mathrm{A}_x$$

he considered that the volumes, being a "sum of areas," would be equally proportional

$$\mathrm{vol(B)} = k\,\mathrm{vol(A)}$$

(with the same constant k), for

$$\mathrm{vol(B)} = \mathrm{omn}\ \mathrm{B}_x = \mathrm{omn}\ k\mathrm{A}_x = k\,\mathrm{vol(A)}.$$

This let Cavalieri deduce the value of an unknown volume from one that he knew. Cavalieri's ideas are akin to those in Archimedes' *Method*, which we will talk about later, but which was unknown at the time. Like several mathematicians of the XVIIth Century (Wallis, Barrow, Pascal, Gregory) Cavalieri came very close to relating integral and derivative. In fact, we can justify Cavalieri's idea of volume as a sum of areas by the same argument with which we proved the Fundamental Theorem of Calculus: say we have a body as in the following picture

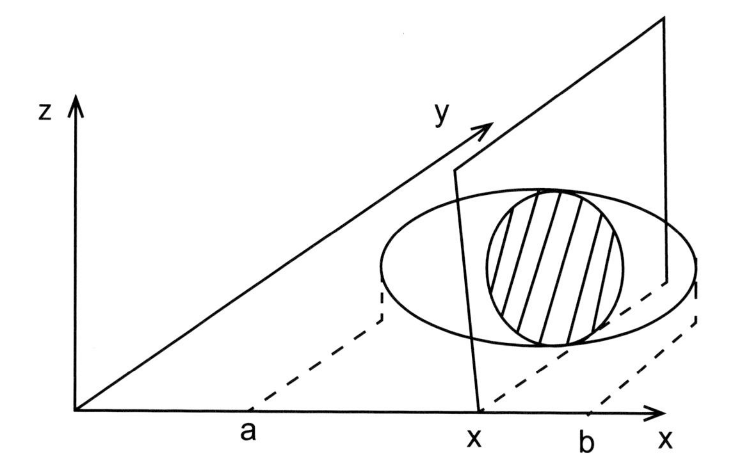

and suppose that for each x between a and b the area of the vertical cut, which we will call $A(x)$, varies continuously. Then the body's volume will be

$$\text{Volume} = \int_a^b A(x)\,dx.$$

To see why, define for each x

$$V(x) = \text{volume of the body from} a \text{to} x.$$

Then $V(x+h) - V(x)$ is the volume of a "slice" of the body, of width h (between x and $x+h$). Then

$$h \min_{x \le t \le x+h} A(t) \le V(x+h) - V(x) \le h \max_{x \le t \le x+h} A(t),$$

so

$$\min_{x \le t \le x+h} A(t) \le \frac{V(x+h) - V(x)}{h} \le \max_{x \le t \le x+h} A(t)$$

and having h tend to zero we have, by the continuity of A,

$$A(x) \le V'(x) \le A(x).$$

In other words, $V' = A$: V is a primitive of A. Therefore

$$\text{Volumen} = V(b) = V(b) - V(a) = \int_a^b A(x)\,dx$$

...something like the sum of areas.

Double Integrals

Volume Under the Graph of $f(x, y)$ Consider a continuous function of two variables, $f : \mathbb{R}^2 \to \mathbb{R}$. Its graph is a surface in 3-space $\mathbb{R}^3$.

$$\text{graph} = \left\{ (x, y, f(x, y)) : (x, y) \in \mathbb{R}^2 \right\}$$

more or less as in the picture

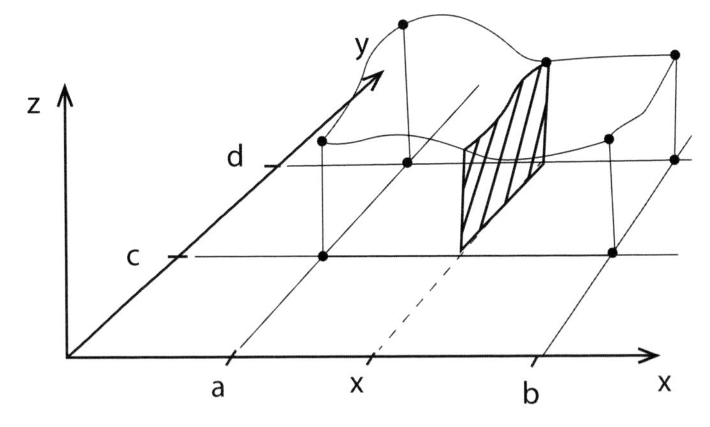

If we want to find the volume under the graph of f (and over the rectangle $[a, b] \times [c, d]$), by Cavalieri, we must calculate

$$\int_a^b A(x)\,dx.$$

But what is $A(x)$? It is the area of the "section" of the body by a vertical plane perpendicular to the x-axis. And the area of this cut is the area under the function with variable y (x is fixed): $f(x, y)$ between c and d, that is:

$$A(x) = \int_c^d f(x, y)\,dy.$$

Thus, the volume under the graph of f over rectangle $[a, b] \times [c, d]$ is

$$\int_a^b \left[\int_c^d f(x, y)\,dy \right] dx,$$

which is a "double integral." It is in fact, two integrals: we first integrate $f(x, y)$ considering that the variable is y (and x is fixed); and then integrate over the variable x. Note that we could repeat the whole argument interchanging the roles of x and y, with sections perpendicular to the y-axis instead of the x-axis. We would then obtain that the volume under the graph of f over rectangle $[a, b] \times [c, d]$ is

$$\int_c^d \left[\int_a^b f(x, y)\, dx \right] dy.$$

Thus we have the equality

$$\int_a^b \int_c^d f(x, y)\, dy\, dx = \int_c^d \int_a^b f(x, y)\, dx\, dy,$$

which is called *Fubini's Theorem.*

Let's do an example: say we want the volume under the graph of $f(x, y) = e^x y^2$ over the rectangle $[0, 1] \times [0, 2]$. Calculate

$$\begin{aligned}\int_0^1 \int_0^2 e^x y^2\, dy\, dx &= \int_0^1 e^x \frac{y^3}{3}\bigg|_0^2 dx \\ &= \int_0^1 e^x \left(\frac{8}{3}\right) dx \\ &= \frac{8}{3} e^x \bigg|_0^1 \\ &= \frac{8}{3}(e - 1).\end{aligned}$$

The Basel Problem

Let's come back to the problem posed by Pietro Mengoli in 1644, and solved by Euler in 1735: calculate the sum of the series $\sum_{k=1}^{\infty} \frac{1}{k^2}$. The proof we give is by Tom Apostol [2].

Basel Problem $\sum_{k=1}^{\infty} \frac{1}{k^2} = \frac{\pi^2}{6}$.

To see this we will calculate—in two different ways—the double integral

$$\int_0^1 \int_0^1 \frac{1}{1 - xy}\, dx\, dy$$

over the square $[0, 1) \times [0, 1)$. The first will give us the left-hand side of the equality, and the second will give us the right-hand side.

The first: since this is an improper integral, we consider

$$\lim_{b\to 1}\int_0^b\int_0^b \frac{1}{1-xy}\,dx\,dy = \lim_{b\to 1}\int_0^b\int_0^b \sum_{k=0}^{\infty} x^k y^k\,dx\,dy,$$

which, since the geometric series converges uniformly in $[-b, b]$, is

$$\begin{aligned}
&= \lim_{b\to 1}\sum_{k=0}^{\infty}\int_0^b\int_0^b x^k y^k\,dx\,dy \\
&= \lim_{b\to 1}\sum_{k=0}^{\infty}\int_0^b x^k dx \int_0^b y^k dy \\
&= \lim_{b\to 1}\sum_{k=0}^{\infty}\frac{x^{k+1}}{k+1}\Big|_0^b \frac{y^{k+1}}{k+1}\Big|_0^b \\
&= \lim_{b\to 1}\sum_{k=0}^{\infty}\left(\frac{b^{k+1}}{k+1}\right)^2 \\
&= \lim_{b\to 1}\sum_{k=1}^{\infty}\frac{b^{2k}}{k^2} \\
&= \sum_{k=1}^{\infty}\frac{1}{k^2},
\end{aligned}$$

for the function $f(x) = \sum_{k=1}^{\infty}\frac{x^{2k}}{k^2}$ is continuous by the Weierstrass M-test: $\frac{x^{2k}}{k^2} \leq \frac{1}{k^2}$, which is summable.

Now we calculate the integral in the second way: we will use different coordinates; u and v as in the figure

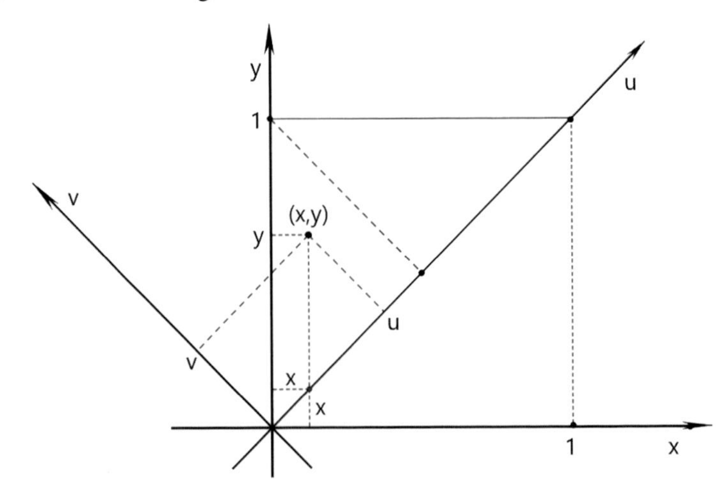

Note also that—by symmetry of $\frac{1}{1-xy}$—the integral over the square is twice the integral over the triangle of vertices $(0, 0)$, $(0, 1)$, $(1, 1)$. We integrate first over the left half of this triangle, and later over the right half. Note that

$$u = \sqrt{2}x + \frac{1}{\sqrt{2}}(y - x) = \frac{1}{\sqrt{2}}(y + x)$$
$$v = \frac{1}{\sqrt{2}}(y - x).$$

Thus, $u + v = \sqrt{2}y$ while $u - v = \sqrt{2}x$, from where $x = \frac{1}{\sqrt{2}}(u - v)$ and $y = \frac{1}{\sqrt{2}}(u + v)$. The function to integrate is then

$$\frac{1}{1 - xy} = \frac{1}{1 - \frac{1}{2}(u^2 - v^2)} = \frac{2}{2 - u^2 + v^2}.$$

A primitive of $\frac{1}{a+v^2}$ is $\frac{1}{\sqrt{a}}\arctan(\frac{v}{\sqrt{a}})$. We integrate over the left half of the triangle:

$$0 \le u \le \frac{\sqrt{2}}{2}$$
$$0 \le v \le u$$

We have

$$\int_0^{\frac{\sqrt{2}}{2}} \int_0^u \frac{2}{2 - u^2 + v^2}\, dv\, du = \int_0^{\frac{\sqrt{2}}{2}} \frac{2}{\sqrt{2 - u^2}} \arctan\left(\frac{v}{\sqrt{2 - u^2}}\right)\Big|_0^u du$$
$$= \int_0^{\frac{\sqrt{2}}{2}} 2\arctan\left(\frac{u}{\sqrt{2 - u^2}}\right) \frac{du}{\sqrt{2 - u^2}} = (*).$$

Now, use the substitution $u = \sqrt{2}\sin t$. Then

$$\frac{u}{\sqrt{2 - u^2}} = \frac{\sqrt{2}\sin t}{\sqrt{2 - 2\sin^2 t}} = \frac{\sqrt{2}\sin t}{\sqrt{2}\cos t} = \frac{\sin t}{\cos t} = \tan t$$
$$\frac{du}{\sqrt{2 - u^2}} = \frac{\sqrt{2}\cos t dt}{\sqrt{2 - 2\sin^2 t}} = \frac{\sqrt{2}\cos t dt}{\sqrt{2}\cos t} = dt$$

so,

$$(*) = \int_0^{\frac{\pi}{6}} 2\arctan(\tan t)\, dt = \int_0^{\frac{\pi}{6}} 2t\, dt = t^2\Big|_0^{\frac{\pi}{6}} = \frac{\pi^2}{36}.$$

Now, the integral over the right half of the triangle:

$$\frac{\sqrt{2}}{2} \leq u \leq \sqrt{2}$$
$$0 \leq v \leq -u + \sqrt{2}.$$

We then have

$$\begin{aligned}\int_{\frac{\sqrt{2}}{2}}^{\sqrt{2}} \int_0^{-u+\sqrt{2}} \frac{2}{2-u^2+v^2}\,dv\,du &= \int_{\frac{\sqrt{2}}{2}}^{\sqrt{2}} \frac{2}{\sqrt{2-u^2}} \arctan\left(\frac{v}{\sqrt{2-u^2}}\right)\Big|_0^{-u+\sqrt{2}} du \\ &= \int_{\frac{\sqrt{2}}{2}}^{\sqrt{2}} 2\arctan\left(\frac{-u+\sqrt{2}}{\sqrt{2-u^2}}\right)\frac{du}{\sqrt{2-u^2}} = (*).\end{aligned}$$

Now use the substitution $u = \sqrt{2}\cos 2t$. Then

$$\frac{-u+\sqrt{2}}{\sqrt{2-u^2}} = \frac{\sqrt{2}(1-\cos 2t)}{\sqrt{2}\sin 2t} = \frac{2\sin^2 t}{2\sin t\cos t} = \tan t$$
$$\frac{du}{\sqrt{2-u^2}} = \frac{-2\sqrt{2}\sin 2t\,dt}{\sqrt{2}\sin 2t} = -2\,dt$$

so,

$$(*) = \int_{\frac{\pi}{6}}^{0} 2\arctan(\tan t)(-2)\,dt = 2\int_0^{\frac{\pi}{6}} 2t\,dt = 2t^2\Big|_0^{\frac{\pi}{6}} = 2\frac{\pi^2}{36}.$$

Finally, the integral is

$$\sum_{k=1}^{\infty}\frac{1}{k^2} = 2\left[\frac{\pi^2}{36} + 2\frac{\pi^2}{36}\right] = \frac{6\pi^2}{36} = \frac{\pi^2}{6}.$$

□

Solids of Revolution

We want to see now how to calculate the volume of a solid of revolution, that is, a body obtained by revolving around an axis, a region enclosed between this axis and a curve:

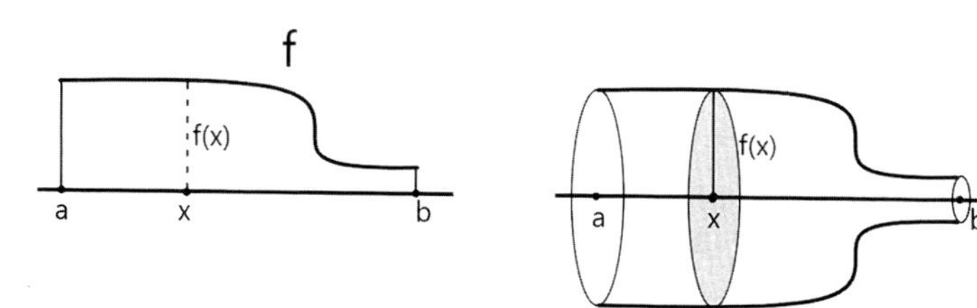

Note that for each x, $A(x)$ is the area of a circle of radius $f(x)$, so the volume of the solid of revolution will be

$$\int_a^b \pi f(x)^2\, dx.$$

Example The volume of a sphere of radius r

We obtain the sphere by revolving the semicircle

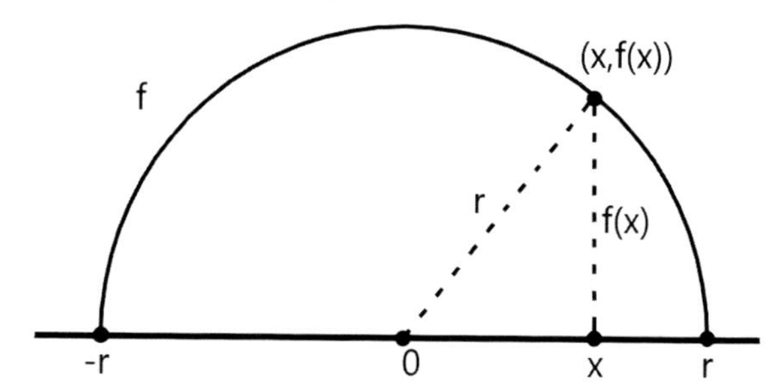

around the x-axis. Here, the distance from $(0, 0)$ to the point $(x, f(x))$ is r, so $x^2 + f(x)^2 = r^2$. Thus, $f(x)^2 = r^2 - x^2$ and the volume of the sphere will be

$$\begin{aligned}
\int_{-r}^{r} \pi f(x)^2\, dx &= \int_{-r}^{r} \pi (r^2 - x^2)\, dx \\
&= \pi \int_{-r}^{r} r^2 - x^2\, dx \\
&= \pi \left(2r^2 r - \frac{x^3}{3}\Big|_{-r}^{r} \right) \\
&= \pi \left(2r^3 - \left[\frac{r^3}{3} - \frac{(-r)^3}{3} \right] \right) \\
&= \pi \left(2r^3 - \frac{2}{3} r^3 \right) \\
&= \pi \left(\frac{6}{3} r^3 - \frac{2}{3} r^3 \right)
\end{aligned}$$

$$= \frac{4}{3}\pi r^3.$$

Example The volume of a torus.

By revolving around the x-axis a circle that does not intersect it, we obtain a *torus*. Its volume will depend on the distance from the axis to the center of the circle, R, and on the radius of the circle, r. We will calculate the volume as the difference between two solids of revolution:

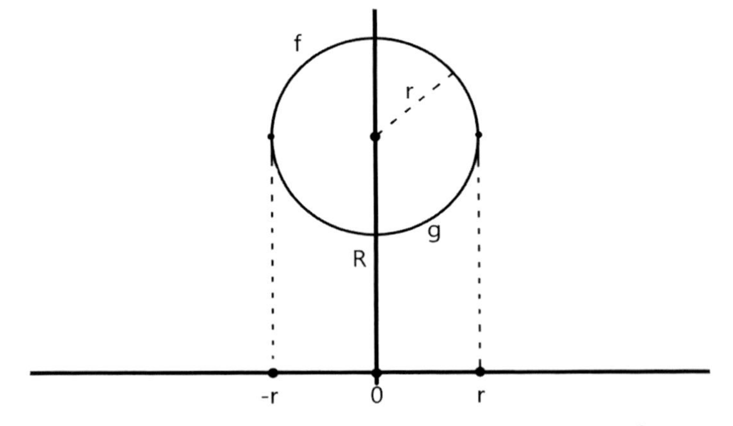

The circumference centered at (0, R) and of radius r has equation $x^2 + (y - R)^2 = r^2$, so

$$y - R = \pm\sqrt{r^2 - x^2}$$
$$y = R \pm \sqrt{r^2 - x^2}.$$

Thus we take $f(x) = R + \sqrt{r^2 - x^2}$ and $g(x) = R - \sqrt{r^2 - x^2}$.

$$\begin{aligned}
\text{Vol} &= \pi \int_{-r}^{r} (R + \sqrt{r^2 - x^2})^2 - (R - \sqrt{r^2 - x^2})^2 \, dx \\
&= \pi \int_{-r}^{r} R^2 + 2R\sqrt{r^2 - x^2} + r^2 - x^2 - R^2 + 2R\sqrt{r^2 - x^2} - r^2 + x^2 \, dx \\
&= 4\pi R \int_{-r}^{r} \sqrt{r^2 - x^2} \, dx \\
&= 4\pi R \cdot \frac{1}{2}(\text{area of the circle centered at } (0, 0) \text{ with radius } r) \\
&= 4\pi R \frac{1}{2}\pi r^2
\end{aligned}$$

$$= 2\pi R\pi r^2.$$

Note that this is the area of the circle of radius r multiplied by the length traveled by its center $(0, R)$ while revolving around the x-axis. We will soon see that this is a special case of a beautiful result known as "Pappus' Theorem."

Example The volume of Gabriel's trumpet.

Evangelista Torricelli (1608–1647) gave, using Cavalieri's principle, the following example, which shows that by revolving a region of infinite area, one may obtain a body of finite volume. Recall that the improper integral $\int_1^\infty \frac{1}{x}\,dx$ is infinite; but now revolve the graph of $f(x) = \frac{1}{x}$ around the x-axis:

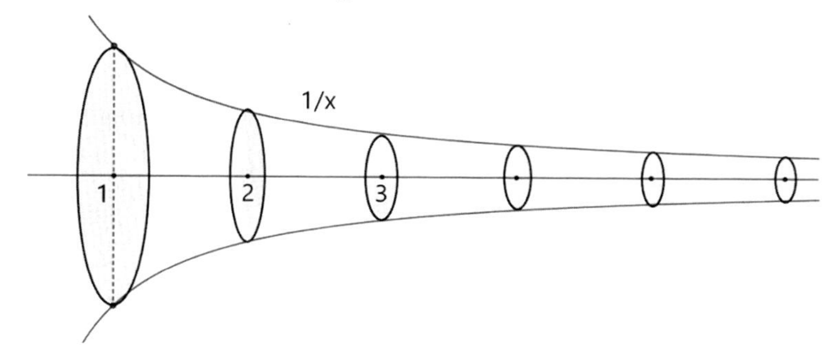

The volume of this solid of revolution is

$$\pi \int_1^\infty \frac{1}{x^2}\,dx = \pi \lim_{b\longrightarrow\infty} \int_1^b \frac{1}{x^2}\,dx = \pi \lim_{b\longrightarrow\infty} \left(-\frac{1}{x}\right)\Big|_1^b = \pi \lim_{b\longrightarrow\infty} \left(-\frac{1}{b}+1\right) = \pi.$$

As you will see in the Exercises, Gabriel's trumpet also has infinite surface area.

Integration of $e^{-\frac{x^2}{2}}$

By using solids of revolution, we will calculate the value of an extremely important integral: the area under the curve that we have called the "Gauss curve" or "bell curve" in Chap. 6. As there is no explicit primitive of $e^{-\frac{x^2}{2}}$, in order to calculate this integral, we will evaluate in two different ways the solid of revolution obtained by making half the Gauss curve rotate around the y-axis:

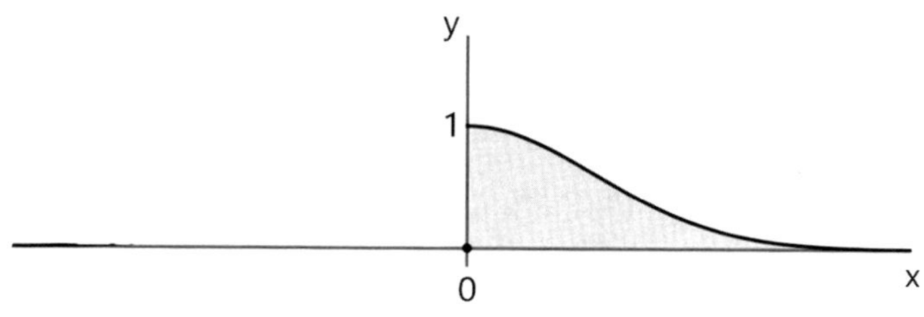

The solid obtained is the region over the (x, z)-plane and under the graph of the function $f : \mathbb{R}^2 \to \mathbb{R}$ given by $f(x, z) = e^{-\frac{x^2+z^2}{2}}$. This volume is

$$\begin{aligned}
\int_{-\infty}^{\infty} \int_{-\infty}^{\infty} e^{-\frac{x^2+z^2}{2}}\, dz\, dx &= \int_{-\infty}^{\infty} \int_{-\infty}^{\infty} e^{-\frac{x^2}{2}} e^{-\frac{z^2}{2}}\, dz\, dx \\
&= \int_{-\infty}^{\infty} e^{-\frac{x^2}{2}}\, dx \int_{-\infty}^{\infty} e^{-\frac{z^2}{2}}\, dz \\
&= I^2,
\end{aligned}$$

where I is the integral we want to evaluate:

$$I = \int_{-\infty}^{\infty} e^{-\frac{x^2}{2}}\, dx.$$

We now consider this region as a solid of revolution. If we are to rotate the curve around the y-axis we must present it as the graph of a function of the variable y. For this, we solve for x as a function of y:

$$\begin{aligned}
y &= e^{-\frac{x^2}{2}} \\
\ln y &= -\frac{x^2}{2} \\
-2\ln y &= x^2 \\
\sqrt{-2\ln y} &= x
\end{aligned}$$

(note that $0 < y \leq 1$, where $\ln y$ is negative). Then the volume will be

$$\begin{aligned}
\pi \int_0^1 -2\ln y\, dy &= -2\pi \lim_{a\to 0} \int_a^1 \ln y\, dy \\
&= -2\pi \lim_{a\to 0} (y\ln y - y)\Big|_a^1 \\
&= -2\pi \lim_{a\to 0} [-1 - a\ln a + a] \\
&= 2\pi
\end{aligned}$$

for, by L'Hôpital, $a \ln a \to 0$. Thus, $I^2 = 2\pi$, and we have

$$\int_{-\infty}^{\infty} e^{-\frac{x^2}{2}}\, dx = \sqrt{2\pi}.$$

Density Functions, Barycenter, and Expectation

In the following sections, two analogous instances of *density functions* will appear, and I believe they deserve a common presentation.

First we will talk about density of mass and barycenter. Suppose we have a rod extending from point a to point b, and set

$$m(x) = \text{ mass of the rod from } a \text{ to } x.$$

Thus, for example, $m(a) = 0$, and $m(b)$ is the total mass of the rod. Then the weight of the rod segment from c to d

a c d b

may be expressed as $m(d) - m(c)$. And if the function $m(x)$ were differentiable (we will suppose that it is), this is

$$\text{mass of the rod from } c \text{ to } d = m(d) - m(c) = \int_c^d m'(x)\,dx.$$

We will write ρ instead of m' and will call this the density function of the rod: a large value of $\rho(x)$ indicates large specific weight close to the point x, while a small $\rho(x)$ indicates low specific weight.

The second instance of density function will appear a few pages later, when we talk about probability. Suppose we have a random variable X (think of this as the possible numerical result of an experiment) that may take values between a and b (we may also have $-\infty < X < \infty$ in some cases). And suppose that for each x between a and b,

$$F(x) = \text{ probability that } a \leq X \leq x.$$

For example, $F(a) = 0$ (we will suppose that there are no isolated values which X attains with positive probability), and $F(b) = 1$, because with probability 1, X will take values between a and b. Just as with the mass m, the probability that X takes a value between c and d may be expressed as $F(d) - F(c)$. And if we suppose that the function F is differentiable, we will have

$$\text{probability that } c \leq X \leq d = F(d) - F(c) = \int_c^d F'(x)\,dx,$$

$F' = f$ is what is known as the probability density function of the variable X.

In both of these instances—density of mass and density of probability—we will define two absolutely analogous notions:

$$\text{the barycenter } c = \frac{1}{m(b)} \int_a^b x\,\rho(x)\,dx,$$

$$\text{and the expectation } \mu = \int_a^b x\, f(x)\, dx.$$

Center of Mass or Barycenter

Suppose we have a rod that goes from point x_1 to point x_2 of the real line, and that it has a weight m_1 at one end and a weight m_2 at the other (and is otherwise massless). At what point will the rod have its center of gravity? In other words, at what point c can it be supported so that it is balanced?

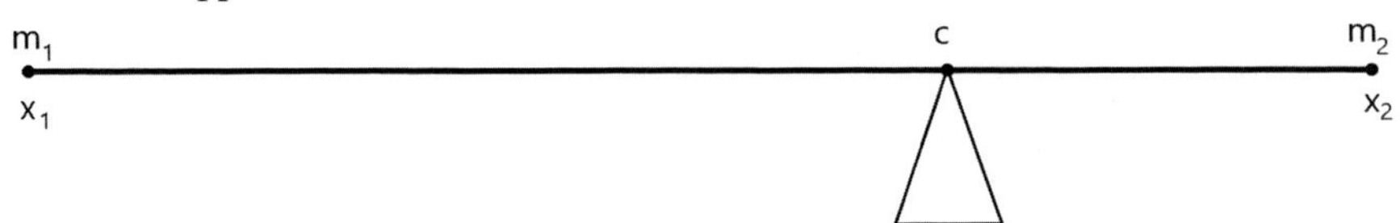

The law of the lever (Archimedes again!) tells us that the moments (force times distance) must be equal. In other words, $m_1(c - x_1) = m_2(x_2 - c)$. From here we obtain the center of gravity:

$$\begin{aligned} m_1(c - x_1) + m_2(c - x_2) &= 0 \\ m_1 c - m_1 x_1 + m_2 c - m_2 x_2 &= 0 \\ (m_1 + m_2)c &= m_1 x_1 + m_2 x_2 \\ c &= \frac{m_1 x_1 + m_2 x_2}{m_1 + m_2}. \end{aligned}$$

So, to have a mass m_1 at x_1 and a mass m_2 at x_2 is like having the total mass $m_1 + m_2$ at c.

Suppose now that instead of having mass concentrated at points x_1 and x_2, we have mass from a to b, of varying density. How do we find the center of mass of the rod? Let's call $c(x)$ the center of mass of the rod from a to x, in other words the point that would be the center of mass if the rod were cut at point x. Also, call $m(x)$ the total mass of the rod from a to x. We need first $c(x + h)$:

c(x) d a x x+h b

Say d is the center of mass of the rod segment between x and $x + h$. Then, just as with the lever, $c(x + h)$ will be

$$c(x + h) = \frac{m(x)c(x) + [m(x + h) - m(x)]d}{m(x + h)}$$

although we do not know d but we do know that $x < d < x + h$. Hence,

$$\frac{m(x)c(x) + [m(x+h) - m(x)]x}{m(x+h)} \leq c(x+h)$$
$$\leq \frac{m(x)c(x) + [m(x+h) - m(x)](x+h)}{m(x+h)}.$$

Then if we multiply by $m(x+h)$ and subtract $m(x)c(x)$,

$$[m(x+h)-m(x)]x \leq m(x+h)c(x+h)-m(x)c(x) \leq [m(x+h)-m(x)](x+h).$$

Dividing by h,

$$\frac{m(x+h) - m(x)}{h}x \leq \frac{m(x+h)c(x+h) - m(x)c(x)}{h} \leq \frac{m(x+h) - m(x)}{h}(x+h)$$

and if we now have h tend to 0,

$$m'(x)x \leq (mc)'(x) \leq m'(x)x.$$

Therefore mc is a primitive of $xm'(x)$. As we have said before $m'(x)$ is the specific weight of the rod at the point x. We call this "density" at x and we denote it by $\rho(x)$. We then have that mc is a primitive of $x\rho(x)$, so

$$m(b)c(b) = m(b)c(b) - \underbrace{m(a)}_{=0}c(a) = \int_a^b x\rho(x)\,dx,$$

and we find the center of mass of the rod

$$c(b) = \frac{1}{m(b)}\int_a^b x\rho(x)\,dx$$

which we may also write

$$c(b) = \frac{\int_a^b x\rho(x)\,dx}{\int_a^b \rho(x)\,dx},$$

where ρ is the density of the rod. Analogously, for a plane region A which has density $\rho(x, y)$, we define its center of mass as

$$c = \frac{\iint_A (x, y)\rho(x, y)\,dx\,dy}{\iint_A \rho(x, y)\,dx\,dy}.$$

Example Center of mass of a cone of height h and radius r.

The center of mass will be on the central axis which we will think of as a rod of length h and variable density $\rho(x)$ proportional to the section area.

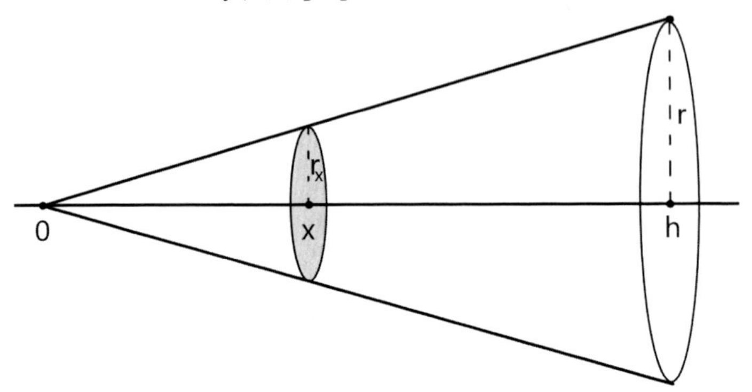

thus, as $\frac{r}{h} = \frac{r_x}{x}$, $r_x = \frac{rx}{h}$ and we set $\rho(x) = k\pi \frac{r^2x^2}{h^2}$. Then

$$c = \frac{\int_0^h k\pi \frac{r^2}{h^2} x^3 \, dx}{\int_0^h k\pi \frac{r^2}{h^2} x^2 \, dx} = \frac{\left.\frac{x^4}{4}\right|_0^h}{\left.\frac{x^3}{3}\right|_0^h} = \frac{3}{4}h.$$

Therefore the barycenter will be at $\frac{1}{4}$ of the height from the base. An interesting fact may be read from our result: the barycenter does not depend on r, in other words it does not depend on the shape of the cone (slender or wider).

Example Stability of a beer can.

Like the Count of Buffon, you also like to have a drink with friends. Let's say that the drink comes in a cylindrical can. Before you start, the can will be full and its barycenter will be at the center. When you have finished it and the can is empty, its barycenter will also be at the center. But it was not always so. When it was half-full, the center of gravity was lower. So the question is: When was the can most stable? In other words, When was its barycenter the lowest possible?

Let's say the height of the can is 1. We will calculate its center of gravity, $c(h)$, when the liquid is at height h $(0 \leq h \leq 1)$, and then we will look for the value of h for $c(h)$ to be minimum. Consider then the can full up to level h as a rod with density

$$\rho = \begin{cases} k, & \text{if } x \leq h \\ 1, & \text{if } x > h, \end{cases}$$

where $k > 1$. Note that when the can is full $(h = 1)$ its weight is

$$\int_0^1 k\,dx = k,$$

but when it's empty ($h = 0$) its weight is

$$\int_0^1 1\,dx = 1.$$

So $k = \frac{\text{weight of the full can}}{\text{weight of the empty can}}$. Fine. When it's full up to level h the barycenter is at

$$\begin{aligned}
c(h) &= \frac{\displaystyle\int_0^1 x\rho(x)\,dx}{\displaystyle\int_0^1 \rho(x)\,dx} \\
&= \frac{\displaystyle\int_0^h kx\,dx + \int_h^1 x\,dx}{\displaystyle\int_0^h k\,dx + \int_h^1 1\,dx} \\
&= \frac{k\frac{x^2}{2}\Big|_0^h + \frac{x^2}{2}\Big|_h^1}{kh + (1-h)} \\
&= \frac{\frac{1}{2}kh^2 + \frac{1}{2} - \frac{h^2}{2}}{(k-1)h + 1} \\
&= \frac{1}{2}\frac{(k-1)h^2 + 1}{(k-1)h + 1}.
\end{aligned}$$

Now we search for the minimum of $c(h)$. Differentiate and set equal to zero

$$c'(h) = \frac{1}{2}\left[\frac{2(k-1)h((k-1)h+1) - ((k-1)h^2+1)(k-1)}{[(k-1)h+1]^2}\right] = 0$$

this happens when

$$\begin{aligned}
0 &= 2(k-1)^2h^2 + 2(k-1)h - (k-1)^2h^2 - (k-1) \\
&= (k-1)[2(k-1)h^2 + 2h - (k-1)h^2 - 1] \\
&= (k-1)[(k-1)h^2 + 2h - 1]
\end{aligned}$$

which is zero for $h = \frac{-2+\sqrt{4+4(k-1)}}{2(k-1)} = \frac{-2+2\sqrt{k}}{2(k-1)} = \frac{\sqrt{k}-1}{k-1}$. For example, if the full can is nine times as heavy as the empty can, this is $h = \frac{1}{4}$.

Pappus' Theorem

Pappus of Alexandria (290–350) was the last of the classical Greek geometers. Let's see one of Pappus' theorems.

Pappus' Theorem *If A is a plane region over the x-axis and S is the solid of revolution obtained by rotating A around the x-axis, then*

$$Vol(S) = area(A) \cdot 2\pi \bar{y},$$

where $\bar{y}$ is the distance from the center of mass of A to the x-axis.

We will do the proof for a region limited above by the graph of a continuous function f, and limited below by the graph of another, g:

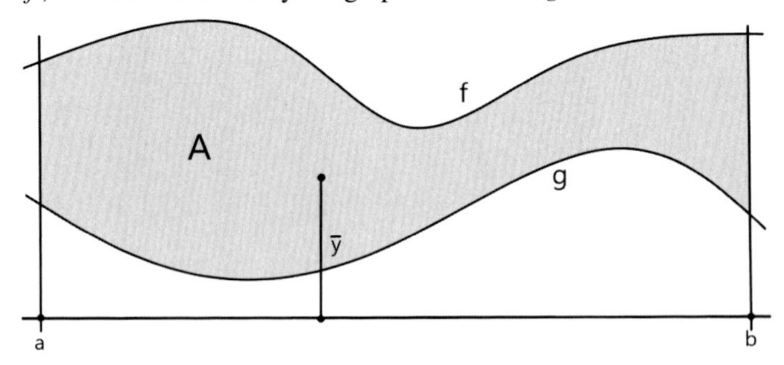

For $\iint_A \rho(x, y)\, dy\, dx = \text{área(A)}$, we set the density as a constant: $\rho = 1$. We calculate $\bar{y}$.

$$\bar{y} = \frac{\int_a^b \int_{g(x)}^{f(x)} y\, dy\, dx}{\iint_A dy\, dx} = \frac{\int_a^b \left. \frac{y^2}{2} \right|_{g(x)}^{f(x)} dx}{\text{area(A)}}$$

$$= \frac{\frac{1}{2} \int_a^b f(x)^2 - g(x)^2\, dx}{\text{area (A)}}$$

$$= \frac{\pi \int_a^b f(x)^2\, dx - \pi \int_a^b g(x)^2\, dx}{2\pi\, \text{area (A)}}$$

$$= \frac{\text{vol(S)}}{2\pi\, \text{area (A)}}.$$

Then

$$\text{vol(S)} = \text{area(A)} \cdot 2\pi\, \bar{y}.$$

□

Note that then the volume of S is the area of A times the length of the circle described by rotating the barycenter of A around the x-axis.

Example The volume of a torus (again!).

We obtain the torus by rotating the circle of center $(0, \mathrm{R})$ and radius r, which clearly has its barycenter at $(0, \mathrm{R})$ and whose area is πr^2. Therefore by Pappus' theorem, the volume of the torus will be

$$\mathrm{vol(torus)} = \pi r^2 \cdot 2\pi \mathrm{R}.$$

Example Center of mass of a semicircle of radius r.

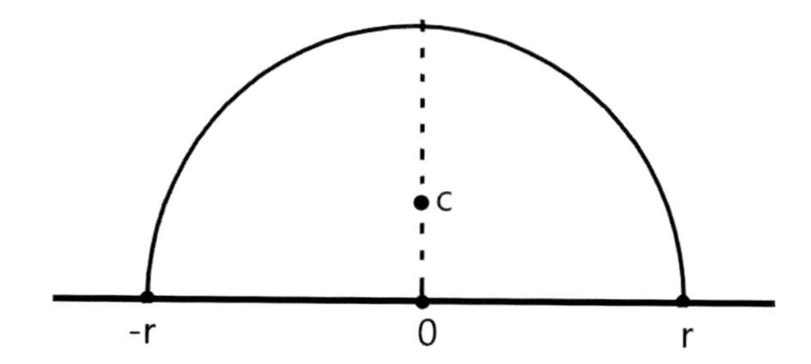

Clearly, by symmetry, the barycenter will be on the y-axis. Say it is the point $c = (0, \bar{y})$. We will calculate $\bar{y}$ using Pappus' theorem. On rotating the semicircle, we obtain a sphere of radius r. Thus its volume is

$$\frac{4}{3}\pi r^3 = \frac{\pi r^2}{2} 2\pi \bar{y},$$

from which we obtain $\bar{y} = \frac{4}{3\pi} r$, and therefore $c = (0, \frac{4}{3\pi} r)$.

Example Barycenter of a triangle T with vertices at $(0, 0)$, $(-1, 0)$ and $(-1, 1)$.

Clearly, the barycenter will be on the median $y = x + 1$. But Pappus will give us the distance to the x-axis: the area of T is $\frac{1}{2}$, and the volume of the cone generated by rotating the triangle around the x-axis is $\frac{1}{3}\pi$. Then

$$2\pi \frac{1}{2} \bar{y} = \frac{1}{3}\pi$$
$$\bar{y} = \frac{1}{3}$$

and for this to be on the median: $\frac{1}{3} = x + 1$, from where $x = -\frac{2}{3}$. Therefore the barycenter we will be at the point $(-\frac{2}{3}, \frac{1}{3})$.

The Method

In 1229, in Jerusalem, a monk scraped and cleaned some old parchments, in order to copy on them a liturgical text. The monk didn't know it, but what he was erasing to make room for the new text was the only remaining copy in the world of Archimedes' *The Method*. The original text of Archimedes (from the IIIrd Century BC) had been transcribed to this parchment in the Xth Century in Constantinople. The old text could barely be seen under the new, but in 1906 a Danish academic studying it discovered Archimedes' barely visible lost work. Since then the *Archimedes palimpsest*, as it is now known, has been submitted to all kinds of image analysis and has been recovered completely.

The Method is a letter from Archimedes to Eratosthenes of Cyrene, a mathematician in the Library of Alexandria:

> Seeing that you are a sincere student ... I have thought it convenient to write to you in detail, explaining a method which will allow you to investigate some mathematical problems through the use of mechanics.

Archimedes then goes on to explain the method which allowed him to guess the value of areas and volumes which he later proved rigorously through exhaustion. Archimedes had discovered the Law of the Lever, and used it to calculate volumes and areas as follows: having a body B of known volume, and wanting to discover the volume of A he would "balance them":

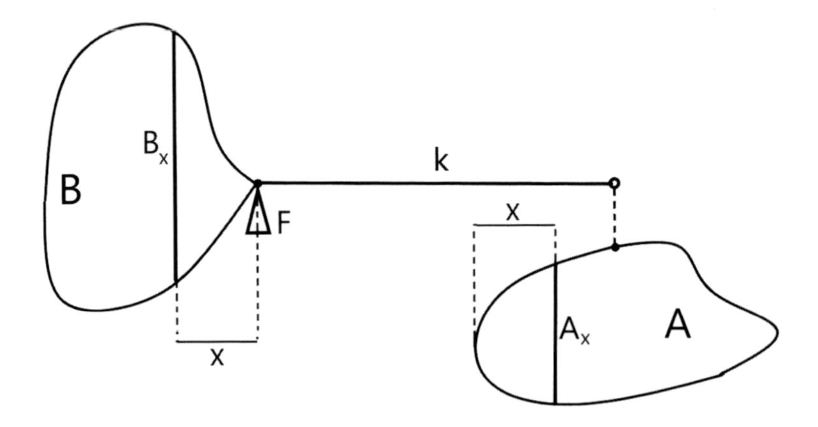

Being of the same width, for each x take sections B_x and A_x; if for each x, B_x at a distance x from the fulcrum F is balanced by A_x at a distance k from the fulcrum, by Archimedes' Law of the Lever

$$xB_x = kA_x \qquad \text{for each } x$$

then the body B times the horizontal distance $\bar{x}$ from its barycenter to F balances the body A at the distance k from F:

$$\bar{x}B = kA$$

from where vol(A) $= \frac{\bar{x}}{k}$vol(B).

But let's do an example. We wish to calculate the area A under $f(x) = x^2$ between 0 and 1. We balance it with the Triangle T of vertices $(0, 0)$, $(-1, 0)$ and $(-1, 1)$ from the last example:

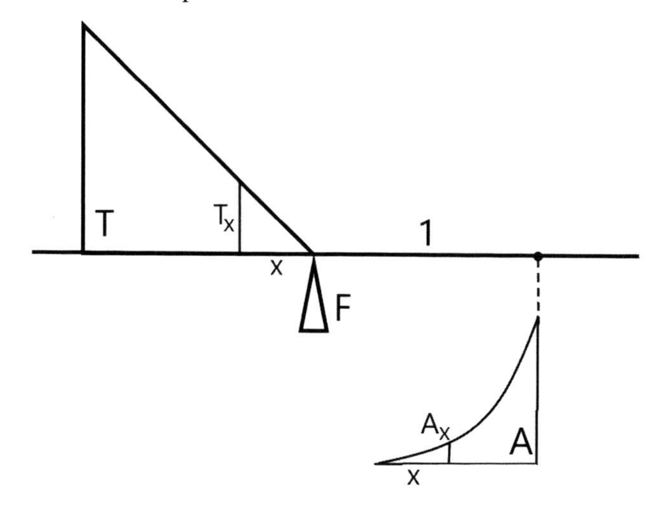

For any given x, we have the relationship between the sections $T_x = x$ and $A_x = x^2$:

$$xT_x = 1 \cdot A_x.$$

Therefore, it must be that $\bar{x}$ · area (T) = area (A):

$$\frac{2}{3} \cdot \frac{1}{2} = \text{area(A)}.$$

In our notation: $\int_0^1 x^2\, dx = \frac{1}{3}$.

But this is almost the same that Cavalieri did! Yes, but Cavalieri did it in 1635... *The Method*, written 2000 years before, had been lost for four centuries and would continue lost another three.

Archimedes of Syracuse (287 a.C.–212 a.C.) was one of the greatest mathematicians. Born in Syracuse (Sicily), he visited Alexandria in his youth, and there met mathematicians with which he maintained contact all his life. He calculated volumes and areas of many figures. He also approximated the number π to an exactness not improved until the XIXth Century. He discovered the Law of the Lever and "Archimedes' principle" on the flotation of bodies. His mechanical devices served to defend his city from the Romans during the Punic Wars.

Surface Area

We will calculate areas of a certain type of surface: the surfaces that correspond to solids of revolution; that is, surfaces which are obtained by revolving a curve around an axis. We need to recall, from elementary geometry, that the surface of a "conical frustum" (see picture)

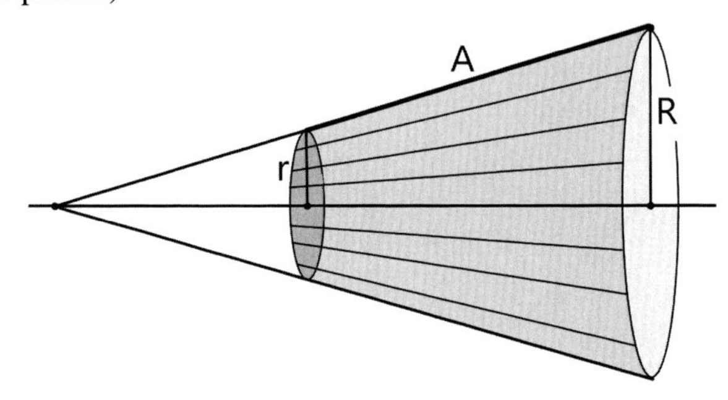

may be written in terms of the radii r and R, and the "slant height" A, as $\pi A(r + R)$. Now consider the surface obtained by revolving the graph of f around the x-axis:

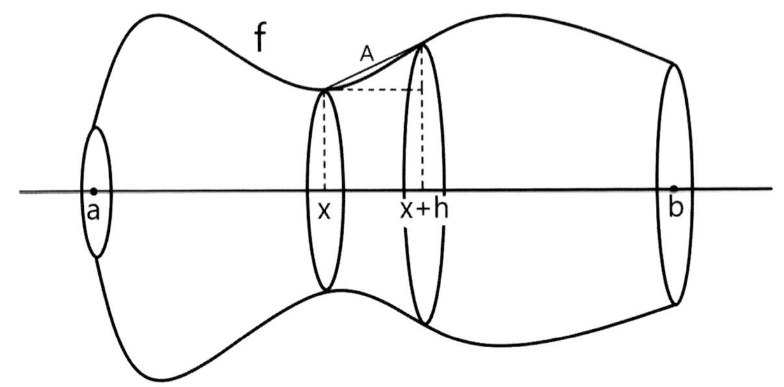

Note first that the slant height A in the picture measures, by Pythagoras' Theorem,

$$A = \sqrt{h^2 + (f(x+h) - f(x))^2}.$$

Call $S(x)$ the area of the surface between a and x, and calculate

$$\begin{aligned}
&S(x+h) - S(x) = \\
&= \text{area between } x \text{ and } x+h \\
&\approx \text{area of the frustum of radii } f(x) \text{ and } f(x+h) \text{ and slant height} A \\
&= \pi A\,(f(x) + f(x+h)) \\
&= \pi \sqrt{h^2 + (f(x+h) - f(x))^2}\,(f(x) + f(x+h))\,.
\end{aligned}$$

Then

$$\frac{S(x+h)-S(x)}{h}=\pi\sqrt{1+\left[\frac{f(x+h)-f(x)}{h}\right]^2}\,(f(x)+f(x+h))$$

which, as $h \to 0$,

$$S'(x)=\pi\sqrt{1+f'(x)^2}\,2f(x).$$

Thus, the surface area between a and b is

$$S(b)=S(b)-S(a)=2\pi\int_a^b f(x)\sqrt{1+f'(x)^2}\,dx.$$

Example Surface of a sphere of radius r.

Revolve, around the x-axis, the semicircumference

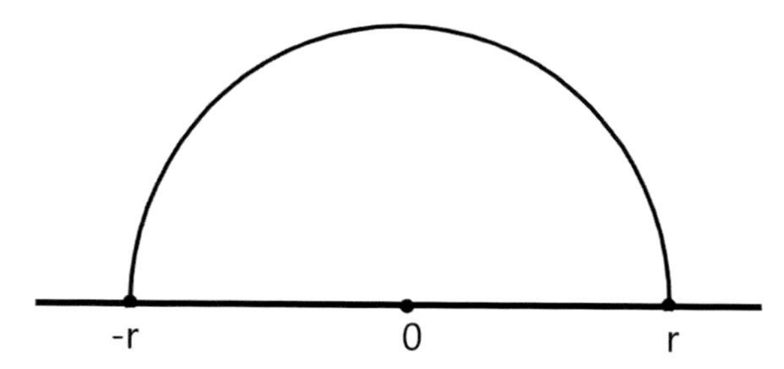

which is the graph of the function $f(x)=\sqrt{r^2-x^2}$. Calculate its derivative:

$$f'(x)=\frac{-x}{\sqrt{r^2-x^2}},$$

thus, the surface area of the sphere is

$$=2\pi\int_{-r}^{r}\sqrt{r^2-x^2}\sqrt{1+\frac{x^2}{r^2-x^2}}\,dx$$

$$=2\pi\int_{-r}^{r}\sqrt{r^2-x^2}\sqrt{\frac{r^2}{r^2-x^2}}\,dx$$

$$=2\pi\int_{-r}^{r} r\,dx=2\pi r(r+r)=4\pi r^2.$$

Note that if we only integrate between a and b we have

$$2\pi r(b-a).$$

Thus, the surface of a "spherical strip" coincides with the surface of the corresponding "cylindrical strip" of radius r:

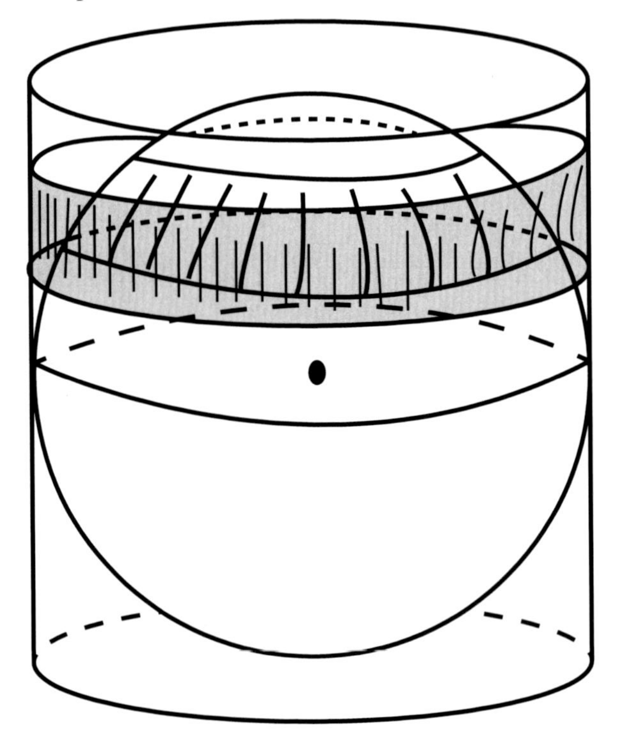

Normal Distribution. Gauss, Laplace, and Stirling

If you practice kicking a ball trying to hit a post, and write down for each shot by how much you missed (-2, if two meters to the left, 1.5 if a meter and a half to the right…) you will have a list of numbers from which you will be able to read these results

(a) Its *mean*. If close to zero, you will have missed to the left and to the right more or less equally. If the mean is positive, you'll know what to do to correct your shots.
(b) You will have an idea of the *dispersion* of your shots. You'll find many $-2, 3, -1, 0, \frac{5}{4}, -\frac{1}{2}, \ldots$, but probably few $28, -37, \ldots$.

To make precise these ideas of mean and dispersion, it is customary to define

$$\text{the mean } \mu = \frac{1}{n}\sum_{i=1}^{n} x_i$$

$$\text{and the standard deviation } \sigma = \sqrt{\frac{1}{n}\sum_{i=1}^{n}(x_i - \mu)^2}.$$

Clearly μ is the mean, and σ indicates how much dispersion there is with respect to this mean: large σ indicates much dispersion; small σ indicates that your shots are all similar. What you probably want in your shots is $\mu = 0$ and small σ.

If you repeated this experiment a very large number of times, you would have a very good idea of the probability of shooting to the left of a point x meters from the post—for any x. Call this probability $F(x)$. And graph its derivative $F' = f$. This is called the density function of your probability, which might look something like this

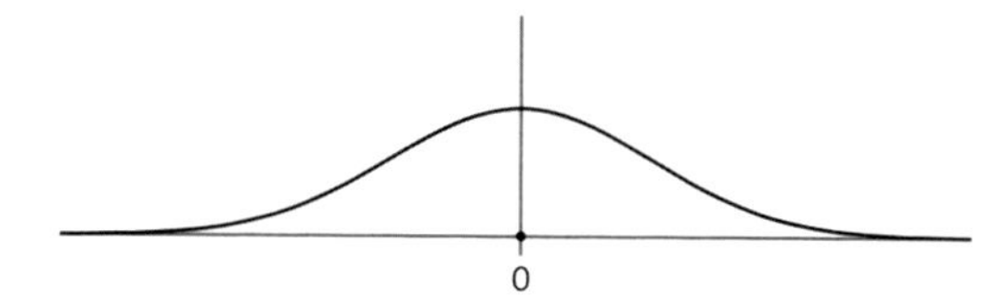

If you now ask yourself: what is the probability of my shot falling between three meters left of the post and two meters to its right (between $x = -3$ and $x = 2$)?, it would be reasonable to compare the area below the curve between -3 and 2 with the total area under the curve (which is one)

$$\frac{\int_{-3}^{2} f}{\int_{-\infty}^{\infty} f} = \int_{-3}^{2} f.$$

In general, given a random variable X, and its probability density function f, the probability that $c \leq X \leq d$ will be

$$\int_{c}^{d} f(x)\,dx.$$

The *expected value* (or mean, or barycenter) of X is

$$\mu = \int_{-\infty}^{\infty} x\, f(x)\,dx,$$

while its *standard deviation* is

$$\sigma = \left(\int_{-\infty}^{\infty} (x - \mu)^2 f(x)\,dx\right)^{\frac{1}{2}}.$$

When we talked about random walk and the Galton board, we discovered Gauss' bell curve: the graph of the function $e^{-\frac{x^2}{2}}$. Later we calculated the area under this curve, which was

$$\int_{-\infty}^{\infty} e^{-\frac{x^2}{2}}\, dx = \sqrt{2\pi}.$$

Thus, the function

$$\frac{1}{\sqrt{2\pi}} e^{-\frac{x^2}{2}}$$

is positive and its integral from $-\infty$ to ∞ is one. This is the density function of the probability distribution called *normal, 0, 1*, (denoted $N(0, 1)$). This distribution has mean $\mu = 0$ and standard deviation $\sigma = \sqrt{1}$. More generally, the *normal,* μ, σ^2 distribution, (denoted $N(\mu, \sigma^2)$), has density function

$$\frac{1}{\sqrt{2\pi\sigma^2}} e^{-\frac{1}{2}\frac{(x-\mu)^2}{\sigma^2}}. \qquad (*)$$

Of course your shots to the post need not have normal distribution. But, like the balls in Galton's board, in the mean, they will. The normal distribution is important because there is a theorem—the Central Limit Theorem—that assures that if a process is repeated many times, a bell curve will appear... Note that I am not giving the statement of the theorem, and we will delve into these matters no further.

To see that the integral over the whole line of the function in $(*)$ is one, calculate

$$\int_{-\infty}^{\infty} e^{-\frac{1}{2}\frac{(x-\mu)^2}{\sigma^2}}\, dx.$$

We use the change of variables $t = \frac{x-\mu}{\sigma}$. Thus,

$$\int_{-\infty}^{\infty} e^{-\frac{1}{2}\frac{(x-\mu)^2}{\sigma^2}}\, dx = \int_{-\infty}^{\infty} e^{-\frac{1}{2}t^2}\sigma\, dt = \sigma\sqrt{2\pi} = \sqrt{2\pi\sigma^2}.$$

In the exercises, you will see that the density function

$$f(x) = \frac{1}{\sqrt{2\pi\sigma^2}} e^{-\frac{1}{2}\frac{(x-\mu)^2}{\sigma^2}},$$

of the $N(\mu, \sigma^2)$ distribution has expected value μ and standard deviation σ. On the other hand, the first derivative of f vanishes on μ, where f reaches a maximum, while its second derivative vanishes at $\mu - \sigma$ and $\mu + \sigma$. Most of the area under f is close to μ, in terms of σ; for example,

$$\int_{\mu-3\sigma}^{\mu+3\sigma} f \approx 0.997\ldots$$

This will help to prove the following.

Laplace's Method *If f has a unique maximum at the point x_0 of (a, b), and $f''(x_0) < 0$, then for large values of n,*

$$\int_a^b e^{nf(x)}\,dx \approx e^{nf(x_0)}\sqrt{\frac{2\pi}{n|f''(x_0)|}}.$$

We give an idea of the proof: approximating f by its Taylor polynomial of order two at x_0, for x near x_0 we have

$$\begin{aligned} f(x) &\approx f(x_0) + f'(x_0)(x - x_0) + \frac{f''(x_0)}{2}(x - x_0)^2 \\ &= f(x_0) + \frac{f''(x_0)}{2}(x - x_0)^2 \\ &= f(x_0) - \frac{|f''(x_0)|(x - x_0)^2}{2} \end{aligned}$$

thus, $nf(x) \approx nf(x_0) - \frac{n|f''(x_0)|(x-x_0)^2}{2}$, and

$$\begin{aligned} \int_a^b e^{nf(x)}\,dx &\approx \int_a^b e^{nf(x_0)} e^{-\frac{n|f''(x_0)|(x-x_0)^2}{2}}\,dx \\ &= e^{nf(x_0)} \int_a^b e^{-\frac{1}{2}\frac{(x-x_0)^2}{\left[\frac{1}{n|f''(x_0)|}\right]}}\,dx \end{aligned}$$

so we are integrating between a and b a normal $N(x_0, \frac{1}{n|f''(x_0)|})$. But for large n the standard deviation σ will be very small, so this is almost like integrating from $-\infty$ to ∞. Then

$$\int_a^b e^{nf(x)}\,dx \approx e^{nf(x_0)} \int_{-\infty}^{\infty} e^{-\frac{1}{2}\frac{(x-x_0)^2}{\left[\frac{1}{n|f''(x_0)|}\right]}}\,dx = e^{nf(x_0)}\sqrt{\frac{2\pi}{n|f''(x_0)|}}.$$

□

As an application, let's see Stirling's formula. Recall the integral formula we saw for $n!$ when studying integration by parts:

$$n! = \int_0^{\infty} e^{-t} t^n\,dt.$$

And let's use Laplace's method to approximate $n!$:

$$
\begin{aligned}
n! &= \int_0^\infty t^n e^{-t}\,dt \quad \text{setting } t = nx, \text{ we have} \\
&= nn^n \int_0^\infty x^n e^{-nx}\,dx \\
&= nn^n \int_0^\infty e^{n\ln x} e^{-nx}\,dx \\
&= nn^n \int_0^\infty e^{n\ln x - nx}\,dx \\
&= nn^n \int_0^\infty e^{n(\ln x - x)}\,dx.
\end{aligned}
$$

Consider then the function $f(x) = \ln x - x$. Its first derivative is $f'(x) = \frac{1}{x} - 1$, which vanishes at $x_0 = 1$; its second derivative is $f''(x) = -\frac{1}{x^2} < 0$. Then, applying Laplace's method,

$$
nn^n \int_0^\infty e^{n(\ln x - x)}\,dx \approx nn^n e^{-n} \sqrt{\frac{2\pi}{n|-1|}} = \frac{n^n}{e^n}\sqrt{\frac{2\pi n^2}{n}} = \frac{n^n}{e^n}\sqrt{2\pi n}.
$$

And we obtain *Stirling's Formula*:

$$
n! \approx \frac{n^n}{e^n}\sqrt{2\pi n}.
$$

Exercises

1 Calculate the volume of a cone of radius r and height h.

2 Calculate the volumes of the solids of revolution obtained by revolving around the x-axis the regions under the following curves.

(a) $y = \sqrt{x}$ (d) $y = \sin x$

(b) $y = x^{1/4}$ (e) $y = \cos x$

(c) $y = x^2$ (f) $y = \sin x + \cos x$.

3 Draw the region $R = \{(x, y) : 0 \le x \le 2, \ \frac{1}{4}x^2 \le y \le 1\}$. Calculate the volume of the solid of revolution obtained by revolving R around

(a) the y-axis,
(b) the line $x = 2$.

4 Calculate the surface area of Gabriel's trumpet.

5 We have a rod on the x-axis, extending from $x = 0$ to $x = L$. Calculate its center of mass if its density is given by:

(a) $\rho(x) = 1$,
(b) $\rho(x) = 1$ between $x = 0$ and $x = L/2$, and $\rho(x) = 2$ between $x = L/2$ and $x = L$,
(c) $\rho(x) = x$,
(d) $\rho(x) = x$ between $x = 0$ and $x = L/2$, and $\rho(x) = L/2$ between $x = L/2$ and $x = L$.

6 For each region R and density ρ, calculate its barycenter

(a) $R = [0, b] \times [0, h]$; with $\rho(x, y) = 1$,

(b) R limited by $y = x^2, x + y = 2$; with $\rho(x, y) = 1$,

(c) R limited by $y^2 = x + 3, y^2 = 5 - x$; with $\rho(x, y) = 1$,

(d) R limited by $y = \sin^2 x, y = 0, 0 \leq x \leq \pi$; with $\rho(x, y) = 1$,

(e) R limited by $y = e^x, y = 0, 0 \leq x \leq a$; with $\rho(x, y) = xy$.

7 Calculate the barycenter of a triangle ABC.

8 Given the density function of the $N(\mu, \sigma^2)$ distribution:

$$\frac{1}{\sqrt{2\pi\sigma^2}} e^{\frac{(x-\mu)^2}{2\sigma^2}},$$

calculate:

(i) its maximum,
(ii) its intervals of growth and decay,
(iii) its intervals of convexity and concavity,
(iv) its expected value,
(v) its standard deviation.

9 The Gamma Function

We end with a presentation of the Gamma function and a few of its properties, including Weierstrass' formula.

The Gamma Function

If one wishes to extend the definition of "$n!$" to other values of the variable (beyond the natural numbers), it makes sense—given the integral expression for $n!$ which we saw in Chap. 5—to attempt

$$x! = \int_0^\infty e^{-t} t^x \, dt.$$

This is of course, an improper integral. Note that for large values of t, $e^{-t}t^x$ is very small, and the function is integrable; while near zero, it is like t^x, and will be integrable for $x > -1$. So the definition makes sense for $x \in (-1, \infty)$. In general, the domain $(0, \infty)$ is preferred, and so x is replaced by $x - 1$. Hence the definition of the *Gamma function*:

$$\Gamma(x) = \int_0^\infty e^{-t} t^{x-1} \, dt.$$

The Γ function is very important in several branches of mathematics, and appears also in many applications. Let's see some of its properties. We know that $\Gamma(1) = 1$ and $\Gamma(n+1) = n!$. Integrating by parts, it is easy to see that

$$\Gamma(x+1) = x\Gamma(x), \qquad \text{for } x > 0.$$

I. Zalduendo, *Calculus off the Beaten Path*, SUMS Readings,
https://doi.org/10.1007/978-3-031-15765-3_9

$\Gamma(0)$ is not defined, but we may use the equality $\Gamma(x) = \frac{\Gamma(x+1)}{x}$ to define Γ in the interval $(-1, 0)$, and then in $(-2, -1)$, $(-3, -2)$... Thus, we consider Γ defined on all real numbers except $\{0, -1, -2, -3, \ldots\}$. One value of Γ which is easy to calculate is $\Gamma(\frac{1}{2}) = \sqrt{\pi}$ (see the Exercises).

Also, Γ is log-convex (and therefore, convex). To see this, if $0 \leq \alpha \leq 1$,

$$\begin{aligned}
\Gamma(\alpha x + (1-\alpha)y) &= \int_0^\infty e^{-t} t^{\alpha x + (1-\alpha)y - \alpha - (1-\alpha)}\, dt \\
&= \int_0^\infty e^{-t\alpha - t(1-\alpha)} t^{\alpha x - \alpha} t^{(1-\alpha)y - (1-\alpha)}\, dt \\
&= \int_0^\infty \left(e^{-t} t^{x-1}\right)^\alpha \left(e^{-t} t^{y-1}\right)^{(1-\alpha)}\, dt \\
&\leq \left(\int_0^\infty e^{-t} t^{x-1}\, dt\right)^\alpha \left(\int_0^\infty e^{-t} t^{y-1}\, dt\right)^{1-\alpha} \\
&= \Gamma(x)^\alpha \Gamma(y)^{1-\alpha},
\end{aligned}$$

(for the inequality, we have used Hölder, from Chap. 7).

The log-convexity together with the equality $\Gamma(x+1) = x\Gamma(x)$ and $\Gamma(1) = 1$ characterize the Γ function.

Weierstrass' Formula

We will now see a formula that presents the Γ function as a product. It will be useful to us in a couple of applications of the Γ function.

Weierstrass' Formula

$$\Gamma(x) = \frac{1}{x} \prod_{k=1}^{\infty} \frac{(1 + \frac{1}{k})^x}{(1 + \frac{x}{k})}.$$

We check it for $0 < x \leq 1$ (it then generalizes to other values through $\Gamma(x+1) = x\Gamma(x)$). Write first

$$\begin{aligned}
\Gamma(n+x+1) &= \Gamma((1-x)(n+1) + x(n+2)) \\
&\leq \Gamma(n+1)^{1-x} \Gamma(n+2)^x \\
&= \Gamma(n+1)^{1-x} (n+1)^x \Gamma(n+1)^x \\
&= (n+1)^x \Gamma(n+1) \\
&= (n+1)^x n! \qquad (9.1)
\end{aligned}$$

On the other hand,

$$\begin{aligned} n! = \Gamma(n+1) &= \Gamma(x(n+x) + (1-x)(n+x+1)) \\ &\leq \Gamma(n+x)^x \Gamma(n+x+1)^{1-x} \\ &= (n+x)^{-x} \Gamma(n+x+1)^x \Gamma(n+x+1)^{1-x} \\ &= (n+x)^{-x} \Gamma(n+x+1). \end{aligned} \tag{9.2}$$

Joining (9.2) and (9.1),

$$(n+x)^x \leq \frac{\Gamma(n+x+1)}{n!} \leq (n+1)^x.$$

Now, dividing by n^x,

$$\left(1+\frac{x}{n}\right)^x \leq \frac{\Gamma(n+x+1)}{n!n^x} \leq \left(1+\frac{1}{n}\right)^x. \tag{9.3}$$

But since

$$\begin{aligned} \Gamma(n+x+1) =& (n+x)\Gamma(n+x) = (n+x)(n+x-1)\Gamma(n+x-1) \\ =& \cdots \\ =& (n+x)(n+x-1)\cdots(x+1)x\Gamma(x), \end{aligned}$$

we may put, in (9.3)

$$\left(1+\frac{x}{n}\right)^x \leq \frac{(n+x)(n+x-1)\cdots(x+1)x}{n!n^x}\Gamma(x) \leq \left(1+\frac{1}{n}\right)^x,$$

and taking $\lim_{n\to\infty}$,

$$1 \leq \lim_{n\to\infty} \frac{(n+x)(n+x-1)\cdots(x+1)x}{n!n^x}\Gamma(x) \leq 1.$$

Then,

$$\Gamma(x) = n^x \frac{n!}{x(x+1)\cdots(x+n)}.$$

Consider the first of the two factors, n^x. Since $\frac{n^x}{(n+1)^x}$ tends to one, we will write $(n+1)^x$ instead of n^x; and with $n+1$ written thus

$$n+1 = \frac{n+1}{n}\frac{n}{n-1}\frac{n-1}{n-2}\cdots\frac{2}{1}$$

$$= \left(1+\frac{1}{n}\right)\left(1+\frac{1}{n-1}\right)\left(1+\frac{1}{n-2}\right)\cdots\left(1+\frac{1}{1}\right)$$
$$= \prod_{k=1}^{n}\left(1+\frac{1}{k}\right),$$

so

$$(n+1)^x = \prod_{k=1}^{n}\left(1+\frac{1}{k}\right)^x.$$

Consider now the second factor.

$$\frac{n!}{x(x+1)(x+2)\cdots(x+n)} = \frac{1}{x}\frac{1}{x+1}\frac{2}{x+2}\cdots\frac{n}{x+n}$$
$$= \frac{1}{x}\prod_{k=1}^{n}\frac{1}{\left(1+\frac{x}{k}\right)}, \text{ for } \frac{k}{x+k} = \left(1+\frac{x}{k}\right)^{-1}.$$

Then,

$$\Gamma(x) = \lim_{n\to\infty}\prod_{k=1}^{n}\left(1+\frac{1}{k}\right)^x \frac{1}{x}\prod_{k=1}^{n}\frac{1}{\left(1+\frac{x}{k}\right)}$$
$$= \frac{1}{x}\lim_{n\to\infty}\prod_{k=1}^{n}\frac{\left(1+\frac{1}{k}\right)^x}{\left(1+\frac{x}{k}\right)}$$
$$= \frac{1}{x}\prod_{k=1}^{\infty}\frac{\left(1+\frac{1}{k}\right)^x}{\left(1+\frac{x}{k}\right)}.$$

□

Let's see a consequence of Weierstrass' formula:

Relation Between $\Gamma(x)$ and $\sin x$ We calculate, using Weierstrass' formula, $\Gamma(x)\Gamma(1-x)$:

$$\Gamma(x)\Gamma(1-x) = \Gamma(x)\Gamma(-x)(-x)$$
$$= \frac{1}{x}\frac{-x}{-x}\prod_{k=1}^{\infty}\frac{\left(1+\frac{1}{k}\right)^x\left(1+\frac{1}{k}\right)^{-x}}{\left(1+\frac{x}{k}\right)\left(1+\frac{-x}{k}\right)}$$
$$= \frac{1}{x}\prod_{k=1}^{\infty}\frac{1}{\left(1-\frac{x^2}{k^2}\right)}.$$

Then,

$$\Gamma(x)\Gamma(1-x)x\prod_{k=1}^{\infty}\left(1-\frac{x^2}{k^2}\right)=1.$$

Note that $x\prod_{k=1}^{\infty}\left(1-\frac{x^2}{k^2}\right)$ vanishes on $x = 0, 1, 2, \ldots$. Euler saw that this was none other than $\frac{\sin(\pi x)}{\pi}$, so that

$$\Gamma(x)\Gamma(1-x)\sin(\pi x)=\pi.$$

If you equate the product expression of $\sin(\pi x)$ to its Taylor series, and look at the coefficient of x^2, you will arrive at the original solution—Euler's—of the Basel problem.

Growth of the Harmonic Series, Again

One last consequence of Weierstrass' formula: we know that

$$\sum_{k=1}^{n}\frac{1}{k}\to\infty$$

as n grows; but we wish to see how fast. To this end, take logarithm in Weierstrass' formula:

$$\ln\Gamma(x)=-\ln x+\sum_{k=1}^{\infty}\left[x\ln\left(1+\frac{1}{k}\right)-\ln\left(1+\frac{x}{k}\right)\right].$$

Differentiating,

$$\begin{aligned}\frac{\Gamma'(x)}{\Gamma(x)}&=-\frac{1}{x}+\sum_{k=1}^{\infty}\left[\ln(1+\frac{1}{k})-\frac{\frac{1}{k}}{(1+\frac{x}{k})}\right]\\&=-\frac{1}{x}+\sum_{k=1}^{\infty}\left[\ln(\frac{k+1}{k})-\frac{1}{k+x}\right]\\&=-\frac{1}{x}+\sum_{k=1}^{\infty}\left[\ln(k+1)-\ln k-\frac{1}{k}+\frac{1}{k}-\frac{1}{k+x}\right].\end{aligned}$$

Now, for $x = 1$:

$$\Gamma'(1) = -1 + \lim_{n\to\infty}\left[\sum_{k=1}^{n}(\ln(k+1) - \ln k) - \sum_{k=1}^{n}\frac{1}{k} + \sum_{k=1}^{n}\left(\frac{1}{k} - \frac{1}{k+1}\right)\right]$$

$$= -1 + \lim_{n\to\infty}\left[\ln(n+1) - \sum_{k=1}^{n}\frac{1}{k} + 1 - \frac{1}{n+1}\right]$$

$$= -1 + \lim_{n\to\infty}\left[\ln(n+1) - \sum_{k=1}^{n}\frac{1}{k}\right] + 1.$$

Finally,

$$-\Gamma'(1) = \lim_{n\to\infty}\left[\sum_{k=1}^{n}\frac{1}{k} - \ln(n+1)\right] = \gamma,\ (Euler's\ constant),$$

which tells us that $\sum_{k=1}^{n}\frac{1}{k}$ grows as $\ln n$.

Exercises

1 Calculate

(i) $\Gamma(7)$.
(ii) $\frac{\Gamma(\frac{16}{3})}{\Gamma(\frac{10}{3})}$.

2 Show that $\Gamma(\frac{1}{2}) = \sqrt{\pi}$. Hint: use the change of variables $u = \sqrt{2t}$.

3 Prove that $\Gamma(x+1) = x\Gamma(x)$ for all $x > 0$.

Bibliography

1. Apostol, Tom, *Calculus vol. 1*, (2nd. Ed.), Blaisdell, Waltham, 1967.
2. Apostol, Tom, A proof that Euler missed: Evaluating $\zeta(2)$ the easy way, Mathematical Intelligencer, 5 (1983), 59–60.
3. Berlinski, David, *A Tour of the Calculus*, Vintage Books, New York, 1997.
4. Boyer, Carl B., *The History of the Calculus and its Conceptual Development*, Dover, New York, 1959.
5. Bressoud, David, *A Radical Approach to Real Analysis* (2nd. Ed.), Math. Assoc. of America, Washington D. C., 2007.
6. Bressoud, David, *Calculus Reordered*, Princeton Univ. Press, Princeton, 2019.
7. Do Carmo, Manfredo P., *Differential Geometry of Curves and Surfaces* (2nd. Ed.), Dover, New York, 2016.
8. Durán, Antonio José, *Historia, con personajes, de los conceptos del cálculo*, Alianza Ed., Madrid, 1996.
9. Fauvel, John and Gray, Jeremy (Eds.), *The History of Mathematics: A Reader*, Macmillan Press, London, 1988.
10. Niven, Ivan, A simple proof that π is irrational, Bull. of the Amer. Math. Soc., vol. 53, no. 6 (1947), 509.
11. Schmidt, Erhard, ber das isoperimetrische Problem im Raum von n Dimensionen, Math. Z. 44, no. 1 (1939), 689–788.
12. Toeplitz, Otto, *The Calculus—A Genetic Approach*, Univ. of Chicago Press, Chicago, 1963.
13. Zalduendo, Ignacio, *Matemática para Iñaki*, Fondo de Cultura Económica, México, 2017.

I. Zalduendo, *Calculus off the Beaten Path*, SUMS Readings,
https://doi.org/10.1007/978-3-031-15765-3

Index

I. Zalduendo, *Calculus off the Beaten Path*, SUMS Readings,
https://doi.org/10.1007/978-3-031-15765-3

M

N

O

P

Q

R

S

Printed in the United States
by Baker & Taylor Publisher Services